GRAVITATIONAL RADIATION AND GRAVITATIONAL COLLAPSE

During the Symposium, S. Chandrasekhar (right) was presented by A. Rubinowicz (left) with the Marian Smoluchowski medal. G. Contopoulos, Secretary General of the IAU, presided over the session at which the ceremony took place.

(Photo by Marek Holzman)

INTERNATIONAL ASTRONOMICAL UNION
UNION ASTRONOMIQUE INTERNATIONALE

SYMPOSIUM No. 64
HELD IN WARSAW, POLAND, 5–8 SEPTEMBER 1973
COPERNICAN SYMPOSIUM

GRAVITATIONAL RADIATION AND GRAVITATIONAL COLLAPSE

EDITED BY

CÉCILE DEWITT-MORETTE

Dept. of Astronomy, University of Texas, Austin, Tex., U.S.A.

D. REIDEL PUBLISHING COMPANY

DORDRECHT-HOLLAND / BOSTON-U.S.A.

1974

Published on behalf of
the International Astronomical Union
by
D. Reidel Publishing Company, P.O. Box 17, Dordrecht, Holland

All Rights Reserved
Copyright © 1974 by the International Astronomical Union

Sold and distributed in the U.S.A., Canada, and Mexico
by D. Reidel Publishing Company, Inc.
306 Dartmouth Street, Boston,
Mass. 02116, U.S.A.

Library of Congress Catalog Card Number 73-91436

Cloth edition: ISBN 90 277 0435 X

Paperback edition: ISBN 90 277 0436 8

No part of this book may be reproduced in any form, by print, photoprint, microfilm, or any other means, without written permission from the publisher

Printed in The Netherlands by D. Reidel, Dordrecht

TABLE OF CONTENTS

PROCEEDINGS OF RECENT MEETINGS ON RELATED SUBJECTS   IX

INTRODUCTION BY C. DEWITT-MORETTE AND A. TRAUTMAN
(*Presented by A. Trautman*)   XI

LIST OF PARTICIPANTS   XIII

PART I / GRAVITATIONAL RADIATION

C. W. MISNER / Mechanisms for the Emission and Absorption of Gravitational Radiation (*Invited Paper*)   3

R. A. MATZNER and Y. NUTKU / The Method of Virtual Quanta and Gravitational Radiation   16

J. A. TYSON / Detection of Gravitational Radiation (*Invited Paper*)   17

V. B. BRAGINSKY / The Prospects for High Sensitivity Gravitational Antennae (*Invited Paper*)   28

Seminar on Experiments Currently in Operation

     M. LEE and J. WEBER / Gravitational Radiation Detector Magnetic Tapes from Rochester and Maryland   35

     A. POVEDA and C. ALLEN / An Upper Limit to the Mass Loss from the Centre of the Galaxy   36

     R. W. P. DREVER, J. HOUGH, R. BLAND, and G. W. LESSNOFF / Observations with Wide-band Gravitational Radiation Detectors   37

     P. KAFKA / On the Evaluation of the Munich-Frascati Weber-Type Experiment   38

     S. BONAZZOLA, M. CHEVRETON, and J. THIERRY-MIEG / Meudon Gravitational Radiation Detection Experiment   39

Seminar on Design of Future Experiments

     S. P. BOUGHN, W. M. FAIRBANK, M. S. MCASHAN, H. J. PAIK, R. C. TABER, T. P. BERNAT, D. G. BLAIR, and W. O. HAMILTON / The Use of Cryogenic Techniques to Achieve High Sensitivity in Gravitational Wave Detectors   40

     D. MAEDER / Optimization of Gravitational Burst Detectors Using Piezoelectric Transducers   52

     D. M. EARDLEY, D. L. LEE, A. P. LIGHTMAN, R. V. WAGONER,

and CLIFFORD M. WILL / Analysis of Gravitational-Wave
Detection Experiments 53
V. B. BRAGINSKY, L. P. GRISHCHUK, A. G. DOROSHKIEVICH,
ya. B. ZEL'DOVICH, I. D. NOVIKOV, and M. V. SAZHIN / Electromagnetic
Detectors of Gravitational Waves (*Invited Paper, presented by Ya. B.
Zel'dovich*) 54
V. DE SABBATA, P. FORTINI, C. GUALDI, and L. FORTINI-BARONI /
Interaction of Gravitational Radiation with a Uniformly Magnetized
Sphere 59
P. J. WESTERVELT / Gravitational Radiation by Ultrarelativistic Bodies 60

PART II / STABILITY AND COLLAPSE

S. CHANDRASEKHAR / The Stability of Relativistic Systems (*Invited Paper*) 63
R. PENROSE / Gravitational Collapse (*Invited Paper*) 82

Seminar on Perturbations and Perturbation Fields Around Black Holes

    S. A. TEUKOLSKY / Perturbations of a Rotating Black Hole 92
    W. H. PRESS / Recent Work on Kerr Stability and Superradiant Wave
    Scattering 93
    J. B. HARTLE / Tidal shifts in Black Holes (*Title Only*)
    A. A. STAROBINSKY / Amplification of Waves Reflected from Kerr
    Black Holes 94
    R. RUFFINI / Focussing and the Focussing Effect of Radiation from
    Ultrarelativistic Orbits (*Title Only*)
    R. RUFFINI and J. ZERILLI / Electromagnetic and Gravitational
    Radiation in Ultrarelativistic Regimes (*Title Only*)
    S. PERSIDES / Scalar Waves in the Exterior of a Schwarzschild Black
    Hole 95
    H. STEPHANI and E. HERLT / Electromagnetic Waves in the Exterior
    of a Schwarzschild Black Hole 96
ya. B. ZEL'DOVICH / Quantum Explosions of White Holes (*Title Only*)
J. FAULKNER / The Role of Gravitational Radiation in the Evolution of
Dwarf Novae 97
L. P. GRISHCHUK, A. G. DOROSHKIEVICH, and Y. YDIN / Gravitational
Waves of Cosmological Wavelength (*Title Only*)
M. A. H. MacCALLUM / On the Description of High-Frequency Gravitational
Waves 98
P. G. BERGMANN / Alternative Approach to Infinity 99
R. W. JOHN / The Geodetic Interval in a Riemannian Space-Time in the
Second Post-Minkowskian Approximation 100
M. A. MELVIN / Magnetization, Matter-Antimatter Symmetry and the
Baryon-Photon Ratio in the Universe 101

A. ROSENBLUM / A New General Covariant Approach to the General
  Relativistic Two-Body Problem   102
T. J. SEJNOWSKI / Gravitational Deviation Reaction   103
T. J. SEJNOWSKI / Tidal Tensor and the Emission and Absorption of
  Gravitational Radiation   104
E. T. NEWMAN / Complex Maxwell and Einstein Fields   105
M. A. MARKOV / On Black and White Holes (*Invited Paper*)   106
J. M. BARDEEN / Properties of Black Holes Relevant to their Observation
  (*Invited Paper*)   132

PART III / ACCRETION OF MATTER AND X-RAY SOURCES

R. GIACCONI / Binary X-ray Sources (*Invited Paper*)   147
J. BREGMAN, D. BUTLER, E. KEMPER, A. KOSKI, R. P. KRAFT, and
  R. P. S. STONE / On the Distance to Cygnus X-1 (HDE 226868)   181

Seminar on the Statistics of Stellar Death

   W. D. ARNETT / Origin of Cosmic Rays, Atomic Nuclei and Pulsars
     in Explosions of Massive Stars   182
   S. VON HOERNER / Some Critical Masses for Gravitational Collapse
     (*Title Only*)

Seminar on Black Holes in Astrophysical Environments

   V. F. SHVARTSMAN / On the Problem of Detection of Isolated Black
     Holes   183
   B. J. CARR and S. W. HAWKING / Black Holes in the Early Universe   184
   S. W. HAWKING and G. W. GIBBONS / Quantum Aspects of Accretion
     onto Black Holes in the Early Universe   185

Seminar on Neutron Stars

   W. KUNDT and H. HEINTZMAN / Pulsar Slowdown and Speedup
     (*Title Only*)
   S. TSURUTA, R. RAMATY, and G. BÖRNER / Surface Composition of
     Neutron Stars that are Accreting Matter   186
   D. M. SEDRAKIAN / The Magnetic Fields of Pulsars   187

Seminar on Exact Solutions in General Relativity

   J. PLEBAŃSKI / A Class of Solutions of Einstein-Maxwell Equations
     with the Cosmological Constant   188
   H. SATO and A. TOMIMATSU / New Solutions of Einstein Equations
     Representing Spinning Masses   191
   L. WITTEN / A New Solution of the Einstein-Maxwell Equations for

a System with Mass, Magnetic Moment, Charge, and Angular momentum 192

R. A. SUNYAEV / Accretion of Matter onto Black Holes (*Invited Paper presented by M. M. Basko*) 193

M. J. REES / Accretion onto Relativistic Objects (*Invited Paper*) 194

L. A. PUSTILNIK and V. F. SHVARTSMAN / On a Possible Influence of Magnetic Fields on the Structure of a Disk Formed During Accretion of Plasma in Binary Systems 213

L. M. OZERNOY / What Information can be Extracted from Radio Data about the Existence of Supermassive Black Holes? 214

P. BOYNTON, J. DEETER, and D. GEREND / Comment on Accretion and Compact X-Ray Source Models 216

SUMMARY BY J. A. WHEELER 217

INDEX OF AUTHORS 224

PROCEEDINGS OF RECENT MEETINGS ON RELATED SUBJECTS

1971 *Proceedings of the International School of Physics 'Enrico Fermi'*, course XLVII (ed. by B. K. Sachs), Academic Press, New York

1972 *Proceedings of the International School of Physics 'Enrico Fermi'* (ed. by B. Bertotti), Academic Press, New York

1972 *Les Houches Summer School 'Black Holes; Astres Occlus'* (ed. by Cécile M. DeWitt and B. S. DeWitt), Gordon and Breach Science Publishers, New York 1973

1972 Texas Symposium on Relativistic Astrophysics
(New York City, December 1972)
Proceedings in press in *Proceedings of the New York Academy of Sciences*

Colloque International du CNRS no. 220,
'Ondes et radiations gravitationnelles', Paris 1973

INTRODUCTION

The IAU Symposium No. 64: *Gravitational Radiation and Gravitational Collapse* took place in Warsaw in conjunction with the Extraordinary General Assembly of the International Astronomical Union held in Poland to commemorate the quincentenary of the birth of Nicolaus Copernicus. It was proper and fitting to organize during the Copernican year a meeting devoted to gravitational phenomena: Copernicus, by putting the Sun at the centre of the planetary system, made the first, decisive step along the road which led to the Newtonian law of universal attraction.

The programme of the meeting was prepared by an Organizing Committee consisting of B. Bertotti, V. B. Braginski, S. Chandrasekhar, C. DeWitt (Editor), W. Fowler, I. D. Novikov, E. Schatzman, K. S. Thorne, A. Trautman (Chairman), J. Weber, J. A. Wheeler, and A. W. Wolfendale. The members of the Committee were asked by the chairman to give their opinion on the format of the meeting, the subjects to be covered and the choice of the invited speakers. Their advice was followed whenever convergent views were expressed, but the chairman is solely responsible for the final programme of the Symposium and its shortcomings. There was a general consensus that the Symposium should concentrate on the discussion of present and future experiments to detect gravitational waves of cosmic origin, and on questions related to collapse, black holes and the stability of relativistic systems. It was clear to all of us that a Symposium such as this could never have taken place in 1973 without the pioneering work of J. Weber.

A spirited discussion among the authors of several experiments on the detection of gravitational radiation was one of the highlights of the meeting. Equally interesting was the evidence for the existence of black holes in the Universe presented by R. Giacconi and supported by several short comments by the participants.

Ya. B. Zel'dovich, although formally not a member of the Organizing Committee, provided us with very valuable advice during the preparation of the Symposium and played a leading role during the meeting itself. K. S. Thorne did probably more than anybody else for the scientific success of the Symposium: with great skill he arranged the entire programme of the afternoon seminars, and, when time was getting short, he resigned from delivering his own report to allow another participant to speak.

On Thursday, September 6, 1973, there was a moving moment for the participants when Wojciech Rubinowicz, the distinguished scientist and President of the Polish Physical Society, presented Subrahmanyan Chandrasekhar with the Marian Smoluchowski medal, awarded to him by the Society in recognition of his outstanding scientific achievements. S. Chandrasekhar developed Smoluchowski's theory of the Brownian motion and applied it to astrophysical problems.

Because of the current importance of Gravitational Radiation and Gravitational Collapse both in physics and in astronomy, the Committee has been urged to make the proceedings available as quickly as possible. For this reason, the proceedings include only the abstracts and references of the contributed papers. The invited papers are published in full; their authors have met the early deadline, often at the cost of great inconvenience. Their cooperation is much appreciated.

The Symposium was supported in part by the Polish Academy of Sciences who provided grants to cover the living expenses of 25 participants of our meeting and offered us free use of many of its facilities. The Executive Committee of the IAU allocated $1500 for travel expenses. The Local Organizing Committee of the Extraordinary General Assembly of the IAU, chaired by J. Smak, did much to make the stay in Warsaw comfortable and pleasant.

CECILE DEWITT-MORETTE

ANDRZEJ TRAUTMAN

LIST OF PARTICIPANTS

Abramowicz, M. A., Zaklad Astronomii PAN, Warszawa, Poland
Aichelburg, P. C., Universität Wien, Wien, Austria
de Angelis, U., Osservatorio Astronomico Capodimonte, Napoli, Italy
Arnett, W. D., Rice University, Houston, U.S.A.
Balkowski, C., Observatoire de Paris, Meudon, France
Bardeen, J. M., Yale University, New Haven, U.S.A.
Basko, M. M., Institute of Applied Mathematics, Moscow, U.S.S.R.
Bergmann, P. G., Syracuse University, Syracuse, U.S.A.
Bertotti, B., Universita di Pavia, Pavia, Italy
Bičak, J., Charles University, Praha, Czechoslovakia
Bludman, S. A., University of Pennsylvania, Philadelphia, U.S.A.
Bonazzola, S., Observatoire de Paris, Meudon, France
Borzeszkowski, H. H., Zentralinstitut für Astrophysik, Potsdam-Babelsberg, G.D.R.
Boynton, P., University of Washington, Seattle, U.S.A.
Braginsky, V. B., Moscow State University, Moscow, U.S.S.R.
Burke, W. L., Lick Observatory, Santa Cruz, U.S.A.
Carr, B. J., University of Cambridge, Cambridge, England
Carson, T. R., University Observatory, Fife, England
Carter, B., University of Cambridge, Cambridge, England
Chandrasekhar, S., University of Chicago, Chicago, U.S.A.
Chevalier, C., Saint-Michel – l'Observatoire, Forcalquier, France
Choquet-Bruhat, Y., Université de Paris, Paris, France
Chrzanowski, P., University of Maryland, College Park, U.S.A.
Chubaryan, E. V., Yerevan State University, Yerevan, U.S.S.R.
O'Connell, R. F., Louisiana State University, Baton Rouge, U.S.A.
Contopoulos, G., Thessaloniki University, Thessaloniki, Greece
Demiański, M., Uniwersytet Warszawski, Warszawa, Poland
DeWitt-Morette, C., University of Texas, Austin, U.S.A.
Doroshkievich, A. G., Institute of Applied Mathematics, Moscow, U.S.S.R.
Doughty, N. A., University of Canterbury, Christchurch, New Zealand
Douglas, D. H., University of Rochester, Rochester, U.S.A.
Drever, R. W. P., University of Glasgow, Scotland
Ehlers, J., Max-Planck-Institut, München, F.R.G.
Fairbank, W. M., Stanford University, Stanford, U.S.A.
Faulkner, J., Lick Observatory, Santa Cruz, U.S.A.
Field, G. B., Harvard College Observatory, Cambridge, U.S.A.
Fishbone, L. G., University of Utah, Salt Lake City, U.S.A.

Forlani, A., Osservatorio Astronomico Capodimonte, Napoli, Italy
Fowler, W. A., California Institute of Technology, Pasadena, U.S.A.
Friedman, J. L., Yale University, New Haven, U.S.A.
Ganea, I. M., Astronomical Observatory, Bucharest, Rumania
Giacconi, R., American Science and Engineering, Cambridge, U.S.A.
Gibbons, G. W., University of Cambridge, Cambridge, England
Godart, O., Université Catholique de Louvain, Louvain, Belgium
Goldberg, L., Kitt Peak National Observatory, Tucson, U.S.A.
Grishchuk, L. P., Sternberg Astronomical Institute, Moscow, U.S.S.R.
Hamilton, W. O., Louisiana State University, Baton Rouge, U.S.A.
Hartle, J. B., University of California, Santa Barbara, U.S.A.
Hawking, S. W., University of Cambridge, Cambridge, England
Heintzmann, H., Universität zu Köln, Köln, F.R.G.
von Hoerner, S., National Radio Observatory, Green Bank, U.S.A.
Imshennik, V. S., Institute of Applied Mathematics, Moscow, U.S.S.R.
Isaacson, R., Illinois Institute of Technology, Chicago, U.S.A.
Ivanenko, D. D., Moscow State University, Moscow, U.S.S.R.
Jaffe, J., Smithsonian Astrophysical Observatory, Cambridge, U.S.A.
John, R. W., Zentralinstitut für Astrophysik, Potsdam-Babelsberg, G.D.R.
Kafka, P., Max-Planck-Institut, München, F.R.G.
Kampfer, B., Friedrich-Schiller-Universität, Jena, G.D.R.
Kardashev, N. S., Sternberg Astronomical Institute, Moscow, U.S.S.R.
Kraft, R. P., Lick Observatory, Santa Cruz, U.S.A.
Kreisel, B., Zentralinstitut für Astrophysik, Potsdam-Babelsberg, G.D.R.
Kundt, W., Universität Bielefeld, Bielefeld, F.R.G.
Langer, J., Charles University, Prague, Czechoslovakia
Lasota, J. P., Zakład Astronomii PAN, Warszawa, Poland
Logan, J. L., Rockefeller University, New York, U.S.A.
Lund, F., University Observatory, Oxford, England
MacCallum, M. A. H., University of Cambridge, Cambridge, England
Markow, M. A., Institute of Nuclear Physics, Moscow, U.S.S.R.
Marx, G., Etvös Fizikai Társulat, Budapest, Hungary
Meader, D., Université de Genève, Genève, Switzerland
Melvin, M. A., Temple University, Philadelphia, U.S.A.
Misner, C. W., University of Maryland, College Park, U.S.A.
Møller, C., NORDITA, København, Denmark
Nadiozhan, P. K., Institute of Applied Mathematics, Moscow, U.S.S.R.
Newman, E. T., University of Pittsburgh, Pittsburgh, U.S.A.
Novikov, I. D., Institute of Applied Mathematics, Moscow, U.S.S.R.
Nutku, Y., Middle Eastern Technical University, Ankara, Turkey
Ozernoy, L. M., Lebedev Institute of Physics, Moscow, U.S.S.R.
Paal, G., Konkoly Observatory, Budapest, Hungary
Paczyński, B., Zakład Astronomii PAN, Warszawa, Poland

Partridge, B., Haverford College, Haverford, U.S.A.
Pathria, R. K., University of Waterloo, Waterloo, Canada
Penrose, R., Oxford University, Oxford, England
Persides, S., University of Thessaloniki, Thessaloniki, Greece
Piotrowski, S., Zakład Astronomii PAN, Warszawa, Poland
Platania, G., Osservatorio Astronomico Capodimonte, Napoli, Italy
Plebański, J. F., Uniwersytet Warszawski, Warszawa, Poland
Poveda, A., Universidad National Autonoma, Mexico City, Mexico
Press, W. H., California Institute of Technology, Pasadena, U.S.A.
Raine, D., University Observatory, Oxford, England
Rees, M., University of Sussex, Brighton, England
Roeder, R. C., University of Toronto, West Hill, Canada
Rosenblum, A., Universität Bonn, Bonn, F.R.G.
Roxburgh, I. W., Queen Mary College, London, England
Ruffini, R., Princeton University, Princeton, U.S.A.
de Sabbata, V., Bologna University, Bologna, Italy
Sato, H., Kyoto University, Kyoto, Japan
Schmutzer, E., Friedrich-Schiller Universität, Jena, G.D.R.
Schramm, D., University of Texas, Austin, U.S.A.
Sedrakian, D. M., Erevan State University, Erevan, U.S.S.R.
Sexl, R., Universität Wien, Wien, Austria
Shvartsman, V. F., Special Astrophysical Observatory, St. Zelenchukskaya, U.S.S.R.
Sida, D. W., Carleton University, Ottawa, Canada
Silk, J., University of California, Berkeley, U.S.A.
Smak, J., Zakład Astronomii PAN, Warszawa, Poland
Sokolov, A. A., Moscow State University, Moscow, U.S.S.R.
Solheim, J. E., Auroral Observatory, Tromsø, Norway
Spyrou, N., University of Thessaloniki, Thessaloniki, Greece
Starobinsky, A. A., Landau Institute for Theoretical Physics, Moscow, U.S.S.R.
Stephani, H., Universitaet Jena, D.D.R.
Stewart, J. M., Max-Planck-Institut, München, F.R.G.
Sugimoto, D., University of Tokyo, Tokyo, Japan
Teitelboim, C., Princeton University, Princeton, U.S.A.
Teukolsky, S., California Institute of Technology, Pasadena, U.S.A.
Thorne, K. S., California Institute of Technology, Pasadena, U.S.A.
Tiomno, J., Institute of Advanced Study, Princeton, U.S.A.
Trautman, A., Uniwersytet Warszawski, Warszawa, Poland
Tsuruta, S., NASA, Greenbelt, U.S.A.
Tyson, J. A., Bell Laboratoires, Murray Hill, U.S.A.
Urbantke, M., Universität Wien, Wien, Austria
Vilain, C., Observatoire de Paris, Meudon, France
Wagoner, R. V., Cornell University, Ithaca, U.S.A.
Walker, M., Max-Planck-Institut, München, F.R.G.

Weber, J., University of Maryland, College Park, U.S.A.
Westervelt, P. J., University of Texas, Austin and Brown University, Providence, U.S.A.
Wheeler, J. A., Princeton University, Princeton, U.S.A.
Witten, L., University of Cincinnati, Cincinnati, U.S.A.
Woodhouse, N. J. M., King's College, London, England
Zel'dovich, Ya. B., Institute of Applied Mathematics, Moscow, U.S.S.R.

PART I

GRAVITATIONAL RADIATION

MECHANISMS FOR THE EMISSION AND ABSORBTION OF GRAVITATIONAL RADIATION*

CHARLES W. MISNER

Dept. of Physics and Astronomy, University of Maryland, College Park, Md. 20742, U.S.A.

Abstract. Following some introductory comments on the fundamentals or first principles governing jointly the emission and absorption of gravitational waves, a list is given of observational targets or goals for gravitational wave astronomy which have been selected from recent critical reviews. Then theoretical studies of plunge radiation and gravitational synchrotron radiation are surveyed, since in this area new techniques are developing rapidly although new observational prospects have not yet been found.

The title of this lecture conveys only a hint of the more precise suggestions of the Organizing Committee. They requested that the talk be "directed towards the future rather than towards reviewing research done in the past" and that it "might contain a synthesis of our theoretical knowledge of the characteristics of the waves bathing the Earth and coming from various sources, so as to provide experimentalists with goals in the design of their future detectors." This is a very demanding assignment, but I can fortunately sidestep the main burden of effort and shorten my own presentation by referring to a number of excellent recent reviews where the desired survey is given more extensively. I will therefore merely state what I take to be the most significant experimental goals, culled from these more detailed surveys, and thereby make room for a few viewpoints which I hope might provoke future theoretical studies in novel directions.

The talk then falls into three parts: first some comments on the fundamentals or first principles governing jointly the emission and absorption of gravitational waves, secondly a list of observational targets or goals for gravitational wave astronomy that I have extracted from recent critical reviews, and thirdly some description of a special area of theoretical studies where new techniques are developing rapidly although they are not currently finding new observational prospects.

1. Fundamental Interactions Between Gravitational Waves and Other Matter

We are all familiar from electromagnetic theory with the fact that good emitters are necessarily good absorbers and *vice-versa*. This requirement on the one hand is necessary to permit thermodynamic equilibrium, and on the other hand follows from the basic microscopic principle of detailed balance. Thus the same fundamentals apply to the study of either emission or detection of gravitational waves, namely the basic interactions between gravitational waves and other fields or matter. It is easiest to begin talking of emitters, but to use linearized theory which is the domain ap-

---

* Supported in part by NASA grant NGR 21-002-010.

plicable to all detectors. A familiar focus in this area of theory is the quadrupole radiation formula for the gravitational wave luminosity of a non-relativistic system:

$$L_{\text{grav. wave}} = \frac{1}{5} \frac{G}{c^5} \langle \dddot{\mathscr{I}}_{jk} \dddot{\mathscr{I}}_{jk} \rangle, \qquad (1)$$

where

$$\mathscr{I}_{jk} = I_{jk} - \tfrac{1}{3} \delta_{jk} I_{ll} \qquad (2)$$

is the trace-free part of the moment of inertia tensor I_{jk} of the system, called the 're-duced quadrupole moment' by MTW (Misner, Thorne, and Wheeler, 1973). More easily applied to astrophysical estimates is the restatement of this formula by Dyson (1962) as

$$L_{\text{grav. wave}} L_0 = (L_{\text{internal}})^2. \qquad (3)$$

Here $L_0 = c^5/G = 3.6 \times 10^{59}$ erg s^{-1} is the natural unit of maximum luminosity, and $L_{\text{internal}} \sim \dddot{\mathscr{I}}_{ij}$ is any asymmetric internal power flow in a system.

Emitters and absorbers which the preceeding formulae suggest are well known, so let me suggest then another viewpoint which may help someone think of still other emitters or absorbers. This viewpoint focusses on anisotropic stresses as the essential feature for interactions with gravitational waves. The linearized field equations for gravitational waves are

$$\Box \bar{h}_{\mu\nu} = 16\pi \, T_{\mu\nu} \qquad (4)$$

but the conservation law $T^{\mu\nu}{}_{,\nu} = 0$ makes the energy-momentum components redundant, and only the spatial components $\Box \bar{h}_{ij} = 16\pi \, T_{ij}$ need be studied. Alternatively one notes that it is the space components h_{ij} of the metric perturbation $h_{\mu\nu} = g_{\mu\nu} - \eta_{\mu\nu}$ which contain the transverse traceless basic wave amplitudes h_{ij}^{TT}. Thus not only is it sufficient to focus on stresses T_{ij} as sources of gravitational waves, but in fact only the anisotropic or traceless part of the stress contributes as a source of h_{ij}^{TT}; simple pressure oscillations are ineffective. For the emission or absorbtion of gravitational waves, then, anisotropic stress is the essential requirement. From this view it is then evident why perfect fluids do not emit, absorb, refract, or reflect gravitational waves. But viscous fluids through viscous shear stress, can absorb gravitational waves (see, e.g. Madore, 1973). Elastic solids in this view are among the most evident candidates for interaction with gravitational waves. With shear stresses T_{ij} found proportional to the wave amplitudes h_{kl} in an elastic solid, one immediately sees that gravitational waves in such a medium move at velocity different from light velocity. They are therefore refracted and partially reflected. But more importantly, energy will be shared between the wave and the solid to make emission and absorbtion possible. Viscosity would then take wave energy absorbed into acoustic oscillations and convert it further into heat.

Unlike the $\mathscr{I}_{jk}$ formula (1), this new centering on anisotropic stresses is not restricted to slowly moving sources, and could therefore lead some insightful person to absorbers or emitters more important than those our limited astrophysical imaginations have so far unearthed. One exotic medium which can support shear stresses is an ideal gas of collisionless particles, such as photons or neutrinos in the early radiation dominated universe. Some properties of gravitational waves moving in this medium have been found by Chesters (1971, 1973), but they would all be more significant at earlier epochs of the universe, beyond the range of validity of his methods, so further thought is in order if one is seriously determined to find possible sources of gravitational waves of greater observational significance than those presently imagined.

Another area that could well be studied further is not just Equation (4), but rather the non-linearities that should be included in even the source-free Einstein wave equation. It would be interesting to find ways in which the frequency or pulse shape of a gravitational wave could change, as occurs in non-linear hydrodynamics where a strong sound wave will build up into a shock wave in time. It is perhaps not adequately investigated in general relativity whether a similar phenomena can occur for strong gravitational waves, producing high frequencies out of an initially low frequency wave. This is important since the largest masses are *a priori* expected to produce very long wavelength gravitational radiation, while shorter wavelengths seem easier to try to observe. The main argument against non-linear frequency multiplication in the Einstein equations is provided by a few examples – the exact solutions for plane and cylindrical gravitational waves. Here, although the full non-linear equations are solved, one basic wave amplitude satisfies a linear equation and no frequency build up is allowed. The needed further studies would then have to consider waves of some more general kind where the Einstein non-linearities are presumably more deep-seated.

We now turn from these few suggestions of where imaginative searches for new sources or detectors might begin, and consider what conclusions have been reached in surveys of now familiar possibilities.

2. Targets for Technical Achievement in Observing Gravitational Waves

Five recent surveys of gravitational waves deserve special attention. Sciama (1972) at the Copenhagen relativity conference in 1971 gave an excellent brief summary of the researches directed by Thorne and by Chandrasekhar which have finally settled theoretical discussions on the existence and sign of gravitational radiation damping. He also discussed the Israel-Carter conjecture to the effect that total gravitational collapse always leads to a Kerr metric black hole, that is to a specific and well studied geometry characterized by nothing but the mass and angular momentum of the collapsed system. In Sciama's review one also finds a description of Hawking's arguments leading to upper bounds of 30% to 50% on the fraction of the initial rest mass energy which could be converted into gravitational waves in a collision and amalgamation of two black holes. These arguments proceed from various lemmas due to Penrose showing that the area of a black hole (i.e. of an event horizon) cannot decrease. One is interested in such exotic processes because only the most extreme conditions conceivable are thought to provide hope for generating gravitational waves at currently detectable intensities.

The Ruffini and Wheeler (1971) article and the Press and Thorne (1972) review were available when Chapters 35-37 of MTW (Misner, Thorne, and Wheeler, 1973) were being written. These chapters provide a textbook introduction and survey of gravitational waves, including possible sources and methods of detection, but Press and Thorne (1971) are more detailed on several points, especially in the analysis of different approaches to detection of gravitational waves. The more recent review Rees gave in Paris at a conference on gravitational waves this June (Rees, 1974) covers the astrophysical aspects of gravitational waves, and goes beyond Press and Thorne in noting a detectability advantage to highly beamed sources even outside our Galaxy, and in considering general energy-type limitations independently of particular source models.

I will not repeat the survey, estimation, and judgement of observability and astrophysical plausibility for all the sources considered in these previous reviews. Rather, I will try to condense and restate some of the conclusions that can be culled from them. But in order to do this I first want to introduce a helpful language. Thus let me define a *Gravitational-wave Pulse Unit* (GPU) as follows:

$$1 \text{ GPU} = 10^5 \text{ erg cm}^{-2} \text{ Hz}^{-1}. \tag{5}$$

This is a measure of the spectral energy density in a gravitational wave pulse at the Earth, so a wave pulse of strength 1 GPU carries 10^5 erg of wave energy across each square centimeter normal to the wavefronts in each 1 Hz frequency interval at the frequency in question. This kind of unit is appropriate when considering pulses whose frequency spectrum is wider than the bandwidth of the detectors, as is expected for current detectors and likely sources. The energy deposited in such a detector is then obtained by multiplying the pulse intensity (erg cm^{-2} Hz^{-1}) by the integral $\int \sigma(\nu) \, d\nu$ of the detector cross section over frequency ('resonance integral'). The size of the unit has been chosen to nearly fit a standard (but hypothetical) 1.8 GPU source to which other sources can be conveniently scaled. The reference source in the following general formula is one which is located at the galactic center and isotropically emits one solar rest mass of energy in a gravitational wave pulse of 1 kHz bandwidth:

$$(\text{pulse intensity}) = 1.8 \text{ GPU} \left(\frac{\text{energy in wave}}{M_\odot c^2}\right) \times$$

$$\times \left(\frac{10 \text{ kpc}}{\text{distance}}\right)^2 \left(\frac{10^3 \text{ Hz}}{\text{bandwidth}}\right)\left(\frac{4\pi}{\Omega}\right). \tag{6}$$

Here Ω is the solid angle into which the source radiates. Using

$$\left(\int \sigma \, d\nu\right)_{\text{Weber}} = 10^{-21} \text{ cm}^2 \text{ Hz} \tag{7}$$

one finds that about 400 GPU are required to dump $1kT$ of energy into a typical Weber bar, and the minimum detectable pulse is determined then by the fraction of kT needed for an acceptable detection efficiency. Rees' (1974) review used the Gibbon-Hawking (1971) estimate which took Weber's limiting sensitivity as about 100 GPU, but Weber (Weber *et al.*, 1973; Lee and Weber, 1974) will suggest substantially lower detectability limits (higher sensitivity) for his present apparatus. Tyson (1973) has used essentially this language in reporting that during a three month period he could exclude the arrival of pulses of strength exceeding 30 GPU which had the frequency (710 Hz) and other characteristics required by his equipment.

Let me now proceed to use this pulse unit to identify several targets or goals of observational capability which can be related to considerations of astrophysical sources. The reviews I have cited discuss many other sources as well, but I have chosen just a few to serve as benchmarks in the rapidly developing science of gravitational wave astronomy.

2.1. Current target

As the current, major, easiest target for a definitive observational consensus I choose the following question: Can one assert the existence or non-existence of pulses of strengths 10^4 GPU or greater arriving at rates of one per month or more? I believe that all current experiments are designed to provide a clear result at at least this level, but unresolved discrepancies between different laboratories at higher sensitivity levels leave the theorist unsure that the valid scope of any observation is clearly understood. Thus a consensus at this target level is expected only in the nearest future.

This immediate target is already of astrophysical interest, because there is a way of saying that it is not excluded by other observations. One can find a pulse strength and rate at these levels by (i) assuming that the source is at the galactic center (as the nearest very large concentration of somewhat mysterious mass). Then (ii) one considers Sciama's argument (1969, see also Field *et al.*, 1969) that a mass loss rate from the center of the Galaxy exceeding 70 $M_\odot$ yr^{-1} during the past 10^9 yr would have disturbed stellar orbits in ways contradicted by observation, and therefore chooses this as a limiting average gravitational wave power. But the waves would be easier to detect if they were infrequent but powerful, rather than frequent and feeble, so one opts for a low pulse rate. One pulse per century would not keep one's interest up, however, so (iii) I have chosen one per month as a low rate that an observer could search for. Then each pulse contains 5 $M_\odot c^2$ of gravitational wave energy, and to again maximize our chances of seeing it, we assume (iv) that it is all beamed within 10^{-3} rad of the galactic plane, which is the smallest angle that remains certain to include the Earth. Then with $(\Omega/4\pi) = 10^{-3}$ and 5 $M_\odot c^2$ energy in Equation (6), and with a bandwidth of 1 kHz as is reasonable for sources in the kHz band where the present detectors are searching, one finds a pulse strength of 10^4 GPU. This target then imagines a source whose properties are not chosen on grounds of astrophysical plausibility, but are designed to specifications set by our humble ignorance. We admit that Divine ingenuity may surpass human in astrophysical model building, and look for the first observation that could test some of the limits.

2.2. Improved Current Target

An event rate of one per month but at a higher sensitivity of 10 GPU is not far from the design capabilities of present detectors, and at this level less extreme astrophysical limits characterize the target. One could obtain these pulses from the same mysterious galactic center source as for the first target, but now it would no longer be necessary to postulate any beaming mechanism. But at this level it is not even necessary to assume a source in the Galaxy. Distant galaxies or primordial cosmological chaos (unknown mechanism in both cases) could be the source of these pulses, as the average energy density in a universal flux of 10 GPU pulses at one per month is 10^{-29} gm cm^{-3}, that is, just within the cosmological limitations set by the Hubble constant for any form of universal energy density.

2.3. High Reliability Target

We now turn to consider observational capabilities well beyond those intended in presently operating instruments. A target which asks little more in sensitivity, but a great deal in reliability, is the ability to see one pulse per year of strength 1 GPU.

To illustrate the severity of these requirements, consider the following experimental specifications (which are probably not optimal) that give a little better than a 90% chance of success after 5 yr of running. Let the detection system have 90% efficiency in finding a gravitational wave pulse of 1 GPU amplitude, and let it report only one false event per decade. Then consider the possible outcomes after it has logged five years of running time, and the significance of each outcome reaching a positive vs a negative conclusion. The 'positive' result finds in favor of the hypothesis that gravitational waves occur with strength $\geqslant 1$ GPU at a rate $\geqslant 1$ per year. The 'negative' result favors the hypothesis of no gravitational wave pulses. The experiment can be considered to have failed if it cannot convincingly distinguish these two hypotheses. This failure occurs if the detection system reports a total of either 2 or 3 events in five years, as 2 favors a negative conclusion by odds of only 15:1, while 3 favors the positive result by only 5:1. But these two awkward outcomes together have less than 10% probability of occuring under either hypothesis, and all other outcomes are decisive by odds greater than 100:1 (actually $\gtrsim 500:1$), so the experiment will be decisive with greater than 90% probability.

These high reliability experiments, although they are substantially more difficult than the experiments aimed at the current targets, are of greater astrophysical significance. The source postulated for the 'current targets' was not expected to exist; it simply could not be excluded by other observations or by energy arguments. For this high reliability experiment target, however, there can be proposed a source model for which there are weak astrophysical suggestions. Thus the observational exclusion of this source, by a negative gravitational wave observational finding, would dispose of what is otherwise a significant theoretical possibility.

The target for high reliability experiments is suggested by the following considerations. Assume there is one star death per year in our Galaxy (estimates run from one to ten). These star deaths could be gravitational collapses which occur without any optical supernova display. It may be that this is the more likely way to produce a black hole, whereas an optical supernova may be associated with the formation of a neutron star as one knows os the case for the crab nebula. The formation of a black hole could release more gravitational wave energy than forming a neutron star, so I have taken an energy relase of 1 $M_\odot c^2$ once a year as conceivable for our Galaxy. This assumes that every case we do not fully understand in the terminal evolution of stars is one where the star converts a substantial fraction of its mass into gravitational waves. Thus a negative observational result at the high reliability target level would show that most star do not end in a blast of gravitational waves, a statement that we cannot make with certainty on other grounds.

2.4. High sensitivity target

This next checkpoint for progress in observational capabilities is motivated by an astrophysical source model which is relatively conservative, i.e. quite definitely expected to exist, namely gravitational waves from neutron star formation at the observed optical supernova rate. The target to which this leads is the capability of detecting pulses of strength 10^{-6} gravitational-wave pulse unit (10^{-6} GPU) occuring at a rate of one per week. The source here would be the Virgo cluster in which there are about 2500 galaxies at distances of 10–15 Mpc. For each of these galaxies one may assume a supernova rate of one SN per 30 yr. The further assumption that each optical supernova leads to the formation of a neutron star is fairly plausible but not undebatable. The energy and bandwidth estimates consider that the oscillation which emits gravitational waves is some strong non-linear vibration of the naescent neutron star. The energy then is a substantial fraction of the binding energy of the neutron star; I take 0.05 $M_\odot c^2$. This is not so much energy that the basic structure of the neutron star is altered, so the vibration frequencies will not be very far from the small oscillation frequencies, and on this ground a bandwidth of 100 Hz is estimated, somewhat narrower than the 10^3 Hz used in estimates for total collapse of stars into black holes. But note that an improved knowledge of neutron star structure, whoch one hopes will become available from studies of pulsars and pulsed X-ray sources, will be needed to select the most reasonable detector frequency within the 500–3000 Hz range to search for these Virgo supernova signals.

The observational targets I have given above are chosen, on the basis of other reviews of a wider range of possibilities, as those most likely to be touched by direct observations along the current line of development. One should also mention, however, possibilities for indirect evidence for the existence of gravitational waves. Later in this symposium Faulkner (1974) will be considering the possibilities for interpreting evolutionary effects in some close binary star systems as governed by gravitational radiation losses. Rees (1974) reviews his suggestion for another indirect effect, namely that primordial gravitational waves with present wavelenths of 1–10 Mpc might lead to apparent failures of the virial theorem for small groups (but not rich clusters) of galaxies by increasing the velocity dispersion of the galaxies.

3. Plunge Radiation and Gravitational Synchrotron Radiation

We now pass on from theory directed immediately to observational prospects, to some recent specialized developments in theoretical methods which, at present, contribute primarily to our basic understanding of gravitational radiation, and to our physical intuitions concerning it. These involve small masses moving in the fields of larger black holes.

Ruffini and Wheeler (1971) made rough estimates of the energy radiated in gravitational waves when a small mass μ plunges radially into a black hole of considerably larger mass M, and verified that this energy is proportional to $(\mu/M)\mu c^2$. The correct coefficient in this formula was then found by Davis et al. (1971), who obtained *

$$E_{\text{grave. wave}} = 0.0104(\mu/M)\mu c^2 \tag{8}$$

using fully relativistic methods developed by Regge and Wheeler (1957) and Zerilli (1970). The only approximation required here, beyond the idealization to point particle motion in a Schwarzschild field, is the assumption that $\mu \ll M$. These techniques were then further applied to circular geodesic motion in a Schwarzschild field by Davis et al. (1972) who found (in the limiting case of an $r=6M$ most tightly bound stable orbit, and to the accuracy I can read their Figure 2) a gravitational wave luminosity of

$$L_{\text{grav. wave}} \sim 10^{-3}(\mu/M)^2 c^5/G \sim$$
$$\sim 200(\mu/M)^2 [M_\odot c^2 \text{ s}^{-1}]. \tag{9}$$

These two calculations can serve as guides in astrophysical estimates, but confirm most importantly, through the μ^2 factor, that small asymmetries are not very effective radiators.

The next set of calculations to be reviewed were part of an attempt (Misner, 1972a) to search for a source model compatible with Weber's (1970) indications of a gravitational wave flux. The aim was to look for particle motions so deep in the gravitational potential well of the black hole that the velocities would be highly relativistic, so the resulting gravitational radiation would have properties analogous to those of electromagnetic synchrotron radiation. These studies did not succeed in producing a plausible model for a source of Weber's gravitational waves, but they did considerably clarify analogies between electromagnetic and gravitational radiation. It now appears that continuing studies may carry the analogy so far that the gravitational theorist may be able to sharpen his intuition by a textbook study of nearly as many model calculations as the electromagnetic theorist has long had available in the standard text chapters on radiation from accelerated charges.

---

* This result has been confirmed by Chung (1973) using entirely independent methods.

The advantages sought in a highly relativistic source were the following. A very large mass would be essential in order to supply the energy to maintain an intense source of gravitational waves at the Galactic center for a substantial period of time such as 10^8 or 10^9 yr. For the efficient production of gravitational waves, strong fields such as those near a black hole appear to be essential. The largest plausible black hole at the center of our own Galaxy would have a mass of the order of $10^8 \, M_\odot$. The highest frequency for particle motion in such a field is about half a cycle per hour. Gravitational radiation at Weber's kilohertz frequencies would therefore have to be a very high harmonic of order $m \sim 10^7$ of the fundamental source frequency, while one knows that highly relativistic charges emit very high harmonics in synchrotron radiation. Another useful feature is that the electromagnetic radiation from a relativistic charge is beamed into a rather narrow angle. We have seen in the discussion of Equation (6) that a beaming factor $\Omega/4\pi = 10^{-3}$ could be employed, putting all the gravitational radiation from the galactic center into a narrow beam near the galactic equator just wide enough to include the Earth to give a corresponding 10^3 reduction in energy requirements. In the first calculations of geodesic synchrotron radiation (Misner et al., 1972), highly relativistic velocities were simply assumed in an unphysical way (not derived from gravitational binding energies) to verify that high harmonics and angular beaming could arise in geodesic motion as well as from accelerated motion. These calculations also were simplified by considering radiation of scalar waves rather than tensor (gravitational) waves. The results showed beaming with $\Omega/4\pi \sim \gamma^{-1}$ where γ is the energy per unit rest mass, as in ordinary synchrotron radiation. The highest harmonics generated, however, were of order $m_{critical} \propto \gamma^2$ in contrast to $m_{critical} \propto \gamma^3$ which holds for accelerated circular orbits.

From this beginning, there were then two lines of development. One pursued the astrophysical prospects for finding a plausible mechanism in which small masses could achieve highly relativistic velocities near black holes so as to get highly beamed, high harmonic, gravitational radiation. This search yielded negative conclusions, and one does not at present see any prospects for finding natural motions which would give rise to such synchrotron-like gravitational radiation. The other line of development was to study in more detail the implications of Einstein's theory for the radiation of gravitational waves by highly relativistic particles.

Let us first consider the search for highly relativistic, astrophysically plausible, sources radiating in synchrotron modes. The first hope was to find GSR (geodesic synchrotron radiation) from stable circular orbits about rapidly rotating black holes, since these orbits have (Boyer-Lindquist) coordinate radii near the horizon and are thus superficially more relativistic than Schwarzschild circular orbits. But already in the first publication (Misner, 1972a) it was known that no GSR is emitted from these orbits. Chrzanowski and Misner had a computation to this effect (reported in Chrzanowski (1972) and Chrzanowski and Misner (1974)), but Bardeen provided the clearer physical explanation, pointing out that the particle motion appeared non-relativistic in a natural (locally non-rotating) reference frame (Bardeen, 1971; Goebel, 1972). For a more complete discussion see Bardeen et al. (1972).

Plunge radiation in the Kerr metric was then considered. When rotation is high, $a \to M$, the small mass may obtain relativistic velocities (as measured in the Bardeen locally non-rotating frame) by the time it reaches the neighborhood of the prograde null circular geodesic where synchrotron modes are most easily excited. One finds, however, that essentially all the radiation is beamed down the black hole and does not excape to infinity. This occurs because, even given angular momenta as high as those of a mass in a stable circular orbit, the infall energy is seen primarily as radial momentum in the Bardeen frame, so the radiation beam is directed radially down the black hole. That his result is plausible can be seen from the work of Bardeen et al. (1972), and definitive calculations from the (scalar) wave equation appear in Chrzanowski's thesis (1973).

Could there be some natural source of gravitational synchrotron radiation which we have been too unimaginative to consider? There are two general lines of argument against this, one from polarization, one from the ineffectiveness of even artificial sources when a natural large parameter $(M-a)^{-1}$ is proposed to give large beaming advantages $(\Omega/4\pi)^{-1}$. An analysis of the response of Weber type antennas to a polarized source (Tyson and Douglass, 1972; Douglass and Tyson, 1972) shows the reported signals could not come from a strongly polarized source at the Galactic center. But we must postulate that the source mechanism is one which is strongly influenced by the plane of the Galaxy in order to obtain beaming in the required direction. It is then natural to assume that this mechanism is able to strongly influence the plane of polarization of the emitted radiation as well. Misner (Misner, 1972b; Misner et al., 1973) and Hughes (Hughes and Misner, 1973) considered the alternative possibility that magnetic-type components of a black hole gravitational field – curvatures generated by the rotation rather than the mass of the black hole – might give gravitational analogs of Faraday rotations. But they find that these effects only influence radiation emitted toward the poles of the black hole (parallel to its rotation) and vanish for radiation in the equatorial plane. Thus no depolarization from gravitational Faraday rotations is to be expected, and the presumption that any highly beamed source would be strongly polarized is strengthened. On this view, then, the observational reports discourage further speculations of a beamed galactic center source (Misner, 1972b).

Another argument against searching further for a beamed galactic center source somehow tied in to a large black hole is the following. Some large parameter of the black hole should control the beaming factor $\Omega/4\pi$. The mass ratio M/μ could be large, but then linear perturbation theory would apply, and the small mass μ then influences only the intensity of the radiation, and not its frequency nor angular distribution. The only other elementary large parameter is $[1-(a/M)]^{-1}$. Bardeen (1974) will describe in another paper at this Symposium why this parameter is unlikely to be really large. But in addition, calculations (Chrzanowski and Misner, 1973) show that even the sources thought to be most effective, although unphysical, get turned off rather than enhanced when this parameter is large, although the physical reasons for this behaviour remain unclear. So for these reasons as well, one cannot plausibly look to a near critical ($a \to M$) rotation velocity of the black hole as the key to synchrotron-like radiation at high harmonics in narrow beams.

Let us now turn to the other line of development from geodesic synchrotron radiation calculations. This is the fuller theoretical understanding of the generation of gravitational waves in new circumstances. Firstly the calculations of synchrotron radiation from energetic ($\gamma \gg 1$) but unstable circular geodesics were elaborated. Chitre and Price (1972) showed that the techniques of spin weighted spherical harmonics and the Newman-Penrose formalism were more efficient for these calculations than the Regge-Wheeler methods used elsewhere, although they were mistaken in interpreting some results as due to failure of the geometrical optics approximation. More detailed descriptions of the vector (electromagnetic) and tensor (gravitational) geodesic synchrotron radiation are given by Breuer et al. (1972), Breuer et al. (1973), and Breuer and Vishveshwara (1973). These calculations all treated Schwarzschild geometry, and provide details of polarization, spectrum, and angular distributions. The geodesic synchrotron radiation is even more strongly polarized than ordinary synchrotron radiation. The frequency spectrum of radiation from these unstable orbits shows a dependence on the spin of the radiation first noted by Davis et al. (1972) at barely relativistic energies ($\gamma < 2.7$). The high γ limit is described by Breuer et al. (1973) who show that, although scalar or vector geodesic synchrotron radiation has the bulk of the power emitted in a single decade of frequency, the corresponding gravitational (tensor) radiation has approximately equal power emitted in each decade from the fundamental up to the cut-off. Near the equatorial plane, however, the energy per square centimeter available for detection is again found mainly in the highest frequency decade, as the beaming factor ($4\pi/\Omega$) is proportional to $\omega^{1/2}$.

To exhibit the physical mechanism responsible for the differing geodesic synchrotron spectra in the scalar, vector, and tensor cases required a further development in computational techniques. Chrzanowski and Misner (1974) developed methods of treating vector and tensor wave equations in a high frequency approximation which were little more involved than methods for scalar wave equations, and allowed GSR calculations to be extended to (unstable) relativistic circular geodesics in the Kerr geometry. These methods emphasized the geometric optics approximation rather than separation of variables. A survey of these techniques is given in Misner (1974). This view allows one to see, through a spacetime integral or inner product $\langle h_{\mu\nu}{}^{TT}, T^{\mu\nu} \rangle$ a frequency dependent factor arising from the angle between the source momentum and the transverse polarization direction of the emitted radiation. This explains the spin dependence of the spectrum.

The computations (Chrzanowski and Misner, (1974) of geodesic synchrotron radiation from relativistic (unstable) circular orbits in the Kerr metric showed one unexpected feature for which no simple physical explanation has yet appeared. This is the cut-off in radiated entensity as $a \to M$. Thus for an orbit of fixed high energy $\mu\gamma$, GSR similar to that from unstable Schwarzschild geodesics is emitted, but the total power and the cut-off frequency vary with a. Both increase slowly as a increases from zero to 0.95 M. But for $a \to M$, the intensity drops sharply to zero while ω_{crit} continues to increase slowly. It therefore appears to be exceptionally difficult to excite synchrotron modes in maximally rotating black holes.

These calculations were much simplified by the circumstance that they were limited to high frequencies with $\omega M \gg 1$, so the wavelength is much smaller than the radius $r \sim M$ of a circular geodesic. The radiation then has as its source a 'near zone' which includes only a small neighborhood of the moving mass μ, but excludes the black hole itself. High frequency radiation emission is thus a much more local phenomenon than quadrupole radiation, so the theorist can gain a deeper understanding of the radiation process and finds continuing analogies to electromagnetism. The most succinct statement fo this local view is that of Matzner and Nutku (1973) who have adapted the Weizsäcker-Williams method to computations of the high frequency part of the spectrum of gravitational radiation from highly relativistic particle motions. Since a report of this work will be given at this Symposium (and since I have also reviewed it elsewhere: Misner, 1974), I will not describe it further here.

Although the technical developments described above have not so far borne fruit in leading us to understand probable sources of gravitational wave pulses, they do seem to carry us forward in narrowing the differences between gravity and electromagnetism. They could, therefore, open up the intuitions of physicists and astrophysicists to as yet unexplored source mechanisms, and thus may be an indirect step toward an understanding of observable gravitational radiation.

References

Bardeen, J. M.: 1971, private communication.
Bardeen, J. M., Press, W. H., and Teukolsky, S. A.: 1972, *Astrophys. J.* **178**, 347.
Bardeen, J. M.: 1974, this volume, p. 132.
Breuer, R. A., Ruffini, R., Tiomno, J., and Vishveshwara, C. V.: 1973, *Phys. Rev.* **D7**, 1002.
Breur, R. A., Tiomno, J., and Vishveshwara, C. V.: 1972, *Lettere Nuovo Cimento* **4**, 857.
Breuer, R. A. and Vishveshwara, C. V.: 1973, *Phys. Rev.* **D7**, 1008.
Chesters, D.: 1971, Ph.D. Thesis, Univ. of Maryland. See abstract and ordering information in *Dissertation Abstracts International* **30**, 3142-B, 1971.
Chesters, D.: 1973, *Phys. Rev.* **D7**, 2863.
Chitre, D. M. and Price, R. H.: 1972, *Phys. Rev. Letters* **29**, 185.
Chrzanowski, P. L.: 1972, *Bull. Am. Phys. Soc.* (II) **17**, 472. (Abstract BJ13).
Chrzanowski, P. L.: 1973, Ph.D. Thesis, University of Maryland, unpublished.
Chrzanowski, P. L. and Misner, C. W.: 1974, 'Geodesic Synchrotron Radiation in the Kerr Geometry by the Method of Asymptotically Factorized Green's Functions', University of Maryland, Center for Theoretical Physics Technical Report, to be published.
Chung, K. P.: 1973, *Nuovo Cimento* **14B**, 293.
Davis, M., Ruffini, R., Press, W. H., and Price, R. H.: 1971, *Phys. Rev. Letters* **27**, 1466; see also Davis, M., Ruffini, R., and Tiomno, J.: 1972, *Phys. Rev.* **D5**, 2932.
Davis, M., Ruffini, R., Tiomno, J., and Zerilli, F.: 1972, *Phys. Rev. Letters* **28**, 1352.
Douglass, D. H. and Tyson, J. A.: 1972, *Astrophys. J.* **178**, 341.
Dyson, F. J.: 1962, unpublished lecture given at Princeton University.
Faulkner, J.: 1974, this volume, p. 97; see also 1971, *Astrophys. J. Letters* **170**, L99.
Field, G. F., Rees, M. J., and Sciama, D. W.: 1969, *Comm. Astrophys. Space Sci.* **1**, 187.
Gibbons, G. W. and Hawking, S. W.: 1971, *Phys. Rev.* **D4**, 2191.
Goebel, C. J.: 1972, *Astrophys. J. Letters* **172**, L95.
Hughes, H. G., III and Misner, C. W.: 1973, 'Gravitational Faraday Rotations of Photons and Gravitons

Propagating in the Kerr Gravitational Field', University of Maryland, Center for Theoretical Physics Report, in preparation.
Lee, M. and Weber, J.: 1974, this volume, p. 35.
Madore, J.: 1973, *Commun. Math. Phys.* **30**, 335.
Matzner, R. A. and Nutku, Y.: 1973, 'On the Method of Virtual Quanta and Gravitational Radiation', University of Maryland Center for Theoretical Physics Report 73-115, *Proc. Roy. Soc.* **A**, in press.
Misner, C. W.: 1972a, *Phys. Rev. Letters* **28**, 994.
Misner, C. W.: 1972b, Lecture at the Sixth Texas Conference on Relativistic Astrophysics, New York, N.Y., December 19, unpublished.
Misner, C. W.: 1974, article in forthcoming volume *Ondes et radiations gravitationelles*, Actes du Colloque International C.N.R.S. No. 220, Paris.
Misner, C. W., Breuer, R. A., Brill, D. R., Chrzanowski, P. L., Hughes, H. G., III, and Pereira, C. M.: 1972, *Phys. Rev. Letters* **28**, 998.
Misner, C. W., Thorne, K. S., and Wheeler, J. A.: 1973, *Gravitation*, Freeman, San Francisco.
Press, W. H. and Thorne, K. S.: 1972, *Ann. Rev. Astron. Astrophys.* **10**, 335.
Rees, M. J.: 1974, article in a forthcoming volume *Ondes et radiations gravitationelles*, Actes du Colloque International C.N.R.S. No. 220, Paris.
Regge, T. and Wheeler, J. A.: 1957, *Phys. Rev.* **108**, 1063.
Ruffini, R. and Wheeler, J. A.: 1971, in A. F. Moore and V. Hardy (eds.), *The Significance of Space Research for Fundamental Physics*, ESRO SP-52, European Space Research Organization, Paris.
Sciama, D. W.: 1969, *Nature* **224**, 1263.
Sciama, D.: 1972, *Gen. Rel. Grav.* **3**, 149.
Tyson, J. A.: 1973, *Physl Rev. Letters* **31**, 326.
Tyson, J. A. and Douglass, D. H.: 1972, *Phys. Rev. Letters* **28**, 991.
Weber, J.: 1970, *Phys. Rev. Letters* **25**, 180.
Weber, J., Lee, M., Gretz, D. J., Rydbeck, G., Trimble, V. L., and Steppel, S.: 1973, *Phys. Rev. Letters* **31**, 779.
Zerilli, F. J.: 1970, *Phys. Rev. Letters* **24**, 737; see also *Phys. Rev.* **D2**, 2141.

THE METHOD OF VIRTUAL QUANTA AND GRAVITATIONAL RADIATION*

RICHARD A. MATZNER

University of Texas, Austin, Tex., U.S.A.

and

YAVUZ NUTKU

Middle Eastern Technical University, Ankara, Turkey

Abstract. We extend the Weizsäcker-Williams method to the domain of gravitational encounters and correlate collision problems with the corresponding interaction of gravitational radiation. To an ultra-relativistic test particle the field of a Schwarzschild mass appears as a pulse of gravitational plane waves. We consider the scattering of each Fourier component, virtual quanta, by the Newtonian-type field of the test body. The scattered flux at infinity gives us the radiative loss of gravitational energy by a rapidly moving particle.

* To be published in *Proc. Roy. Soc. London.*

DETECTION OF GRAVITATIONAL RADIATION

J. A. TYSON

Bell Laboratories, Murray Hill, N.J. 07974, U.S.A.

Abstract. The claims by J. Weber must be examined in the context of system time-response and signal-to-noise ratio. After briefly discussing the sources and spectral distribution of system noise, we discuss optimal time-domain filtering. Although five independent groups have searched for two-antenna coincidences, only Weber has claimed any excess at zero time lag. We present preliminary results of a high sensitivity two-antenna coincidence search.

In this talk I will attempt to review the problem of maximizing the sensitivity of Weber-type gravitational antennas, and I will give a summary of the latest searches for gravitational radiation (GR). The techniques of GR detection have reached such a point that the next generation of antennas may be sufficiently sensitive to GR that we can envision all sky surveys and searches for GR from likely astrophysical sources. In this new field we are always reminded of the pioneering efforts of Joseph Weber. Without his continuing work we would not be here today discussing GR as a possible experimental reality.

Weber's claimed observation (Weber 1969, 1970a, b, 1972) of intensive bursts of kilohertz-band GR is based on two significant features in his data: (1) an excess number of coincidences above chance between two antennas, and (2) a sidereal correlation of these coincidence events. I will discuss here only tests to check on his claim (1) of a statistically significant excess at zero time lag. After reviewing the detection sensitivity limits set by extraneous system noise, I will discuss the limits presently set by several independent experiments, with emphasis on a two-antenna coincidence experiment at Bell Labs and the University of Rochester.

As the gravitational wave passes by an elastic solid, the wave does work against the electrical forces in the solid and, in the case of a Weber-type antenna, the aluminum bar absorbs some energy from the wave. This energy absorption cross section is proportional to the mass of the bar multiplied by $(v/c)^2$, where v is the sound velocity in the bar. Obviously, we could significantly improve the absorption cross section by fabricating the antenna from solid nuclear matter. But adequately instrumenting a neutron star would seem an impossible task. More exactly, for a geometrically linear antenna the gravitational wave appears as an acceleration gradient along the axis of the antenna, thereby coupling only to the odd longitudinal elastic modes of oscillation. For a bar geometry and in the case of a broadband source spectrum (a short burst of GR), the integrated absorption cross section is given by

$$\int \upsilon(v)\,dv = 2Gm\omega^2 l^2/\pi^3 c^3 n^2,$$

where we have substituted the dispersion relation for a bar of length l, longitudinal resonance frequency ω, and mode number n. This absorbed energy will produce an observable change in the bar energy if the obscuring noise is comparatively small. We must study how the various sources of extraneous noise may be minimized.

We wish to convert the bar oscillations into an electrical signal, adding as little noise as possible. If the elastic potential energy in an odd longitudinal mode of the bar is measured by attaching electromechanical transducers symmetrically around the center of the bar, the electrical equivalent circuit appears as in Figure 1. The narrow mechanical resonance of the bar is represented by the electrical resonance of $L_1 C_1 R_1$. The electromechanical transducer of capacitance C_2 electromechanical efficiency

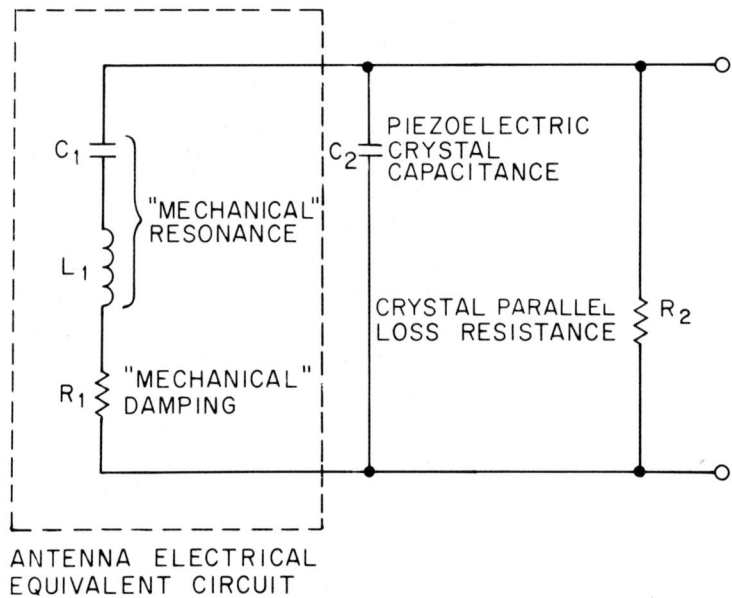

Fig. 1. The equivalent electrical circuit of the antenna, looking into the transducer terminals. The voltage available is a fraction $\beta = C_1/C_2$ of the voltage across C_1, representing the electromechanical efficiency β.

$\beta = C_1/C_2$, and parallel loss R_2 creates the electrical impedance of the antenna as seen across the two terminals on the right. The narrowband noise due to the Johnson noise of R_1 filtering through the antenna impedance is the Brownian motion of the bar in this single mode. Small changes in the bar energy are masked by both this Brownian motion and also the wideband noise from the transducer loss R_2. In addition to these two fundamental noise sources, we must contend with the additional series and parallel noises of the preamplifier, as shown in Figure 2. The wideband noise from R_s is the limiting noise in all the present experiments searching for GR. This is because we expect a sudden change in bar energy upon reception of a burst of GR, and therefore we must have a wide electronics bandwidth in order to optimize detection of this kind of signature.

The spectral distribution of noise coming from the antenna and first amplifier is sketched in Figure 3. The occurrance of a detected GR burst would appear as a brief

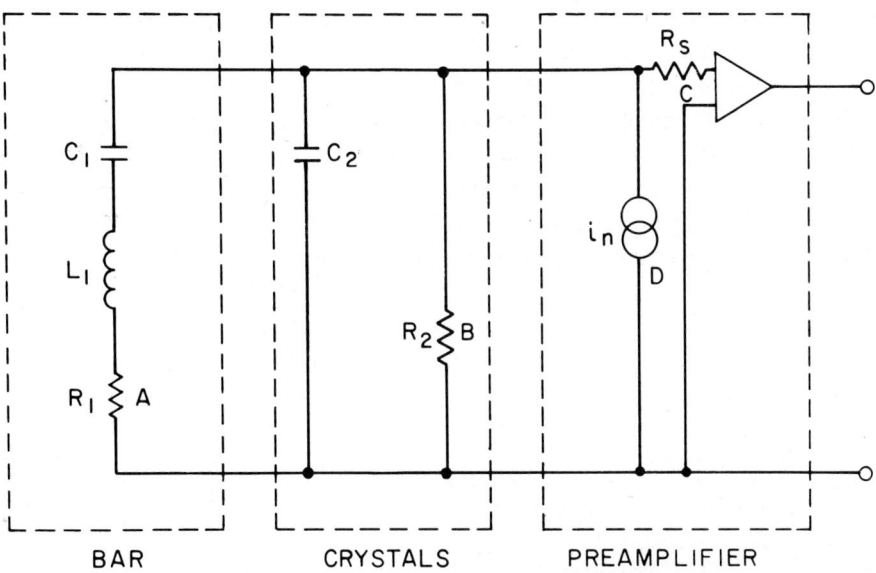

Fig. 2. The four irreducible noise sources (R_1, R_2, R_s, i_n) are shown in this equivalent input circuit.

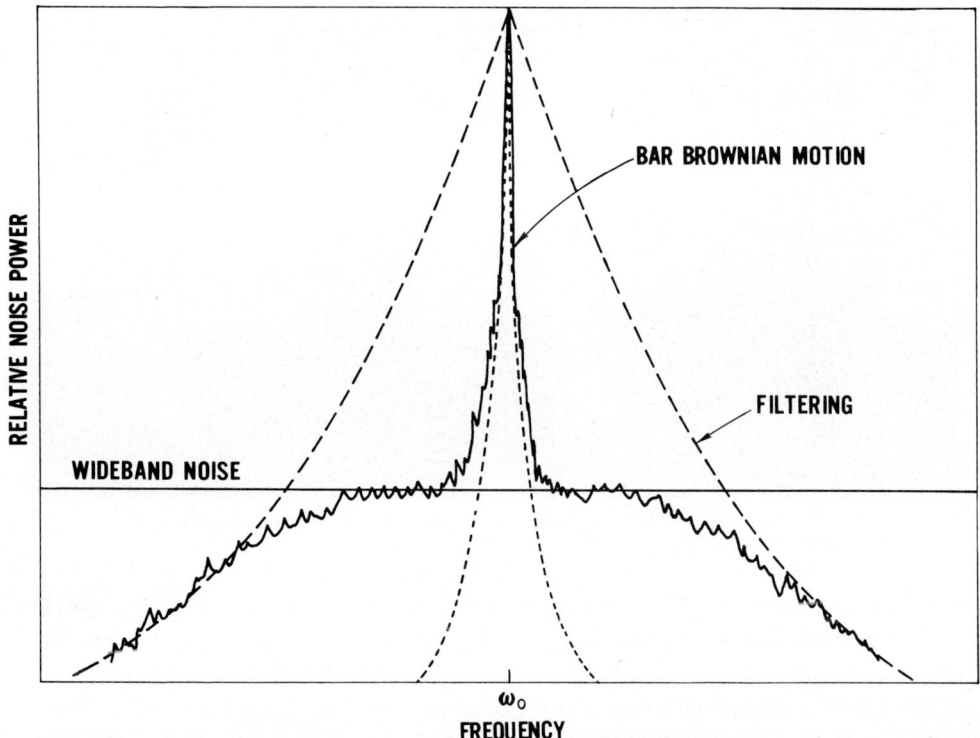

Fig. 3. A sketch of the spectral distribution of the antenna noise near the longitudinal resonance. The synchronous detector (demodulator) translates the peak in the spectrum to zero frequency.

increase in sideband power and a relatively permanent change in power at resonance over a time equal to the Brownian motion autocorrelation time. The nonzero bandwidth of the Brownian motion power is due to unavoidable mechanical losses in the aluminum, rather than GR losses. Postdetection electrical low pass filtering eliminates wideband noise components at sideband frequencies higher than the Nyquist frequency corresponding to the data sampling rate. In most of the present GR antennas, the wideband noise plateau is more than a thousand times less than the Brownian motion power at the center of the Lorentzian line resonance.

The new GR antenna at Bell Labs is shown in Figure 4. The main transducers are mounted symmetrically about the center of the 3.6×10^6 gm, 375 cm long aluminum bar. This symmetry cancels any sensitivity to bending modes which can couple to microseisms. Other modifications of the Weber detector include transducers on the

Fig. 4. Photograph of the new Bell Labs antenna with vacuum tank disassembled.

end, and low frequency air flotation of the entire vacuum tank assembly. The 710 Hz resonance of the bar is synchronously detected, and the 0.1 sec averaged components (X, Y) of the instantaneous antenna vector in the complex plane are sampled at a 10 Hz rate by a clock-tape data system. The amplitude of the bar oscillation (distance from the origin in the complex plane) is shown as a function of time in Figure 5. The slow change is Brownian motion of the bar (autocorrelation time 100 s), whereas the fast, small amplitude noise is due primarily to the preamplifier. With this system, the total extraneous wideband noise power is sixteen times smaller than the Brownian motion power.

We wish to maximize our sensitivity to sudden changes in bar energy. Figure 6 shows the assumed signal signature and the expression for the total power signal to noise ratio for the detection of this signature. The noise denominator contains contributions from rapid components of the Brownian motion (proportional to the time resolution divided by the bar autocorrelation time), the preamplifier noises, and the transducer loss tangent, $\tan \delta$. If the preamplifier offset noise R_s dominates the transducer loss noise, the maximum signal to noise ratio (S/N) as a function of time resolu-

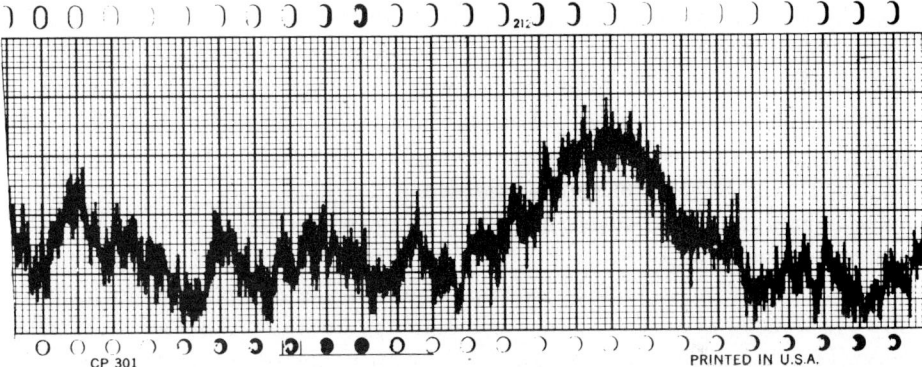

Fig. 5. Output amplitude of the Bell Labs antenna after demodulation. Slow changes are kT noise from the bar, and fast noise is the preamplifier noise. Time scale: 30 s per small division.

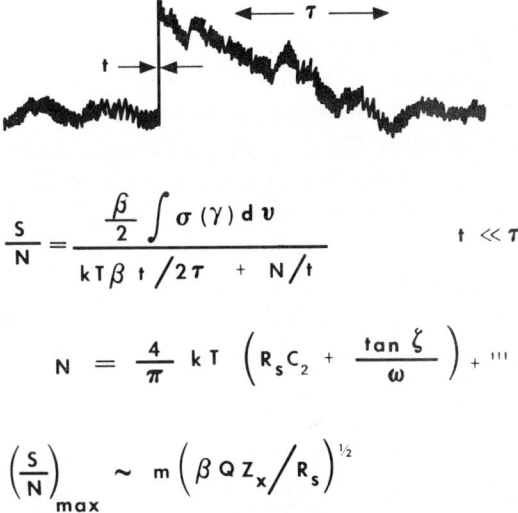

$$\frac{S}{N} = \frac{\frac{\beta}{2} \int \sigma(\gamma) d\upsilon}{kT\beta\, t/2\tau \;+\; N/t} \qquad t \ll \tau$$

$$N = \frac{4}{\pi} kT \left(R_s C_2 + \frac{\tan \zeta}{\omega} \right) + \cdots$$

$$\left(\frac{S}{N}\right)_{max} \sim m \left(\beta Q Z_x / R_s \right)^{1/2}$$

Fig. 6. Output amplitude signature due to a short burst of GR. If the wideband noises from the transducer loss and the preamplifier parallel noise are less than the preamplifier series noise from R_s, then the final expression for the power signal to noise ratio is obtained.

tion is proportional to $m(\beta Q/N)^{1/2}$, where $Q = \omega\tau/2$. We have maximized this for the new Bell Labs and U. Rochester antennas. There is some confusion regarding this notation. A $\frac{1}{4}kT$ event in our antenna would correspond to a $\frac{1}{40}kT$ event in an antenna of $\frac{1}{10}$ the mass and the same $\beta Q/N$. Quoting an event in fractions of kT (the thermal noise of that antenna) does not imply a unique GR flux, unless $\beta Q/N$ and m are known. A more sensitive way to search for GR is to look in the complex plane for a sudden change on the antenna vector. This is done in our computer program.

The number distribution for these sudden changes is shown as a function of square amplitude for 0.1 s time resolution in Figure 7. For 1 s resolution (longer integration), fewer than one $\frac{1}{2}kT$ (in the Bell Labs antenna) event per month would be expected for this system. We have carried out a search for three months and have found no events larger than this. This null result (Tyson, 1973) shows that the flux of GR at 710 Hz is considerably smaller now than Weber's claimed flux at 1660 Hz in 1970. A $\frac{1}{2}kT$ excitation in the Bell Labs antenna would correspond to a GR flux which, if present at 1660 Hz, would give Weber's antenna of 1970 an excitation of less than $\frac{1}{200}kT$. Because of occasional local interference, a more sensitive search may be carried out between two antennas in coincidence.

Several groups are now searching with two antennas in coincidence. The relative sensitivity assuming equal electronics noise is shown in Figure 8. These data are from preprints kindly supplied to us by the various groups. The electronics noise factor N varies over a factor of ten among the various groups, but this technology is changing rapidly. (The factor $(\beta Q/N)^{1/2}$ represents the relative intensity of Brownian motion over wideband noise. The factor of bar mass m comes from the integrated absorption cross section.) Note the large increase in sensitivity of Weber's experiment between 1970 and present. The present values of the electronics noise N put Munich, Frascati, Paris, and Maryland roughly equal at the best sensitivity at ~ 1660 Hz, Moscow somewhat lower, and Bell Labs and U. Rochester nearly equal at 710 Hz at a sensitivity twice as high as any antennas at 1660 Hz. If we had the combination of the Paris preamp and the Bell Labs bar and transducer, we would have a sensitivity six times higher. All the groups are presently improving their signal to noise, and we expect to see an order of magnitude increase in sensitivity within the next year. All these groups have so far indicated a null result for their search for GR, except Weber's group.

One possible explanation of this disparity may be in the kind of signature which is searched for. Most of the groups search in coincidence for a more or less sudden change in bar amplitude or phase, whereas Weber's current experiment searches in coincidence for fixed threshold crossings of a type of time derivative of the antenna vector: $\dot{X}^2 + \dot{Y}^2$. Any *momentary* departure in relative phase or amplitude would yield a spike in this squared derivative, but would not be detected in the linear searches for GR performed by these other groups. In fact, Weber's best results (highest value of excess coincidences at zero time lag) seem to have been with nonlinear detection schemes. Since Weber's signal to noise ratio (real to random rate) has remained relatively constant during a greater than ten fold increase in sensitivity to GR, this suggests some other origin for these coincidences in the nonlinear detector. However, these important questions must be settled by direct experimentation. J. Weber has performed tests and has offered his help in the resolution of this problem. In collaboration with D. H. Douglass (University of Rochester) we are now recording X, Y, and $\dot{X}^2 + \dot{Y}^2$ on digital magnetic tape at both the Bell Labs and Rochester antennas.

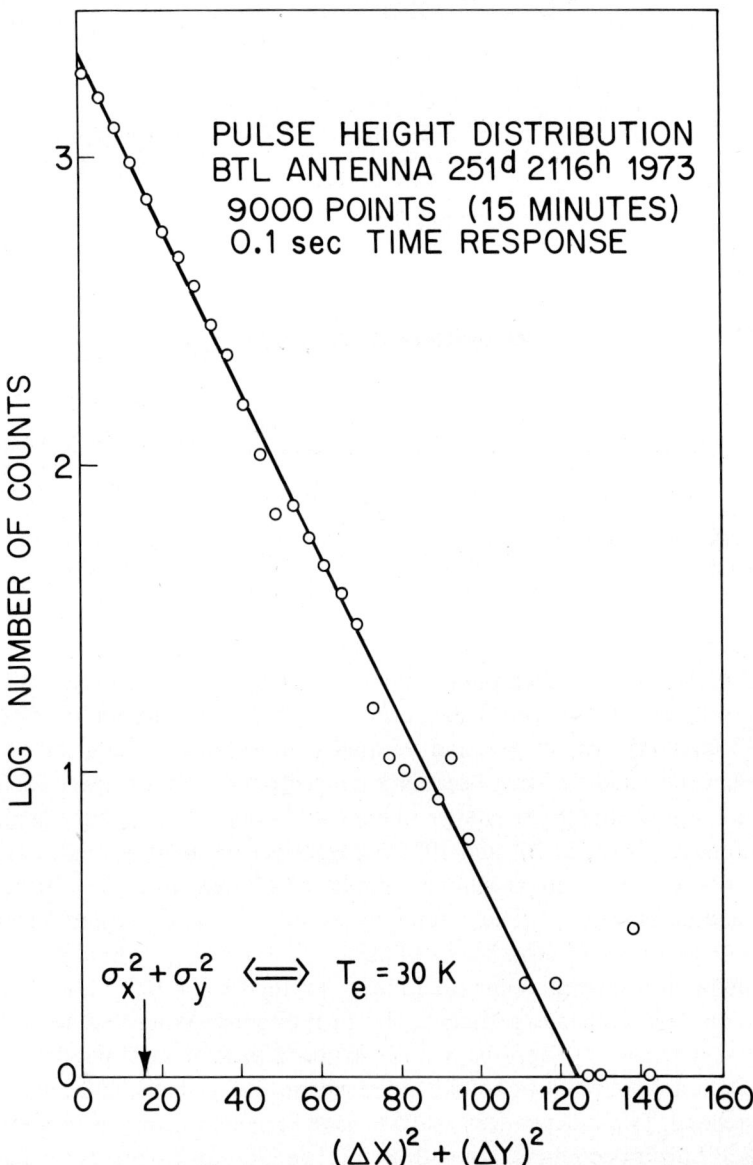

Fig. 7. Number distribution of sudden (<0.1 s) changes in amplitude or phase. Amplifier noise dominates.

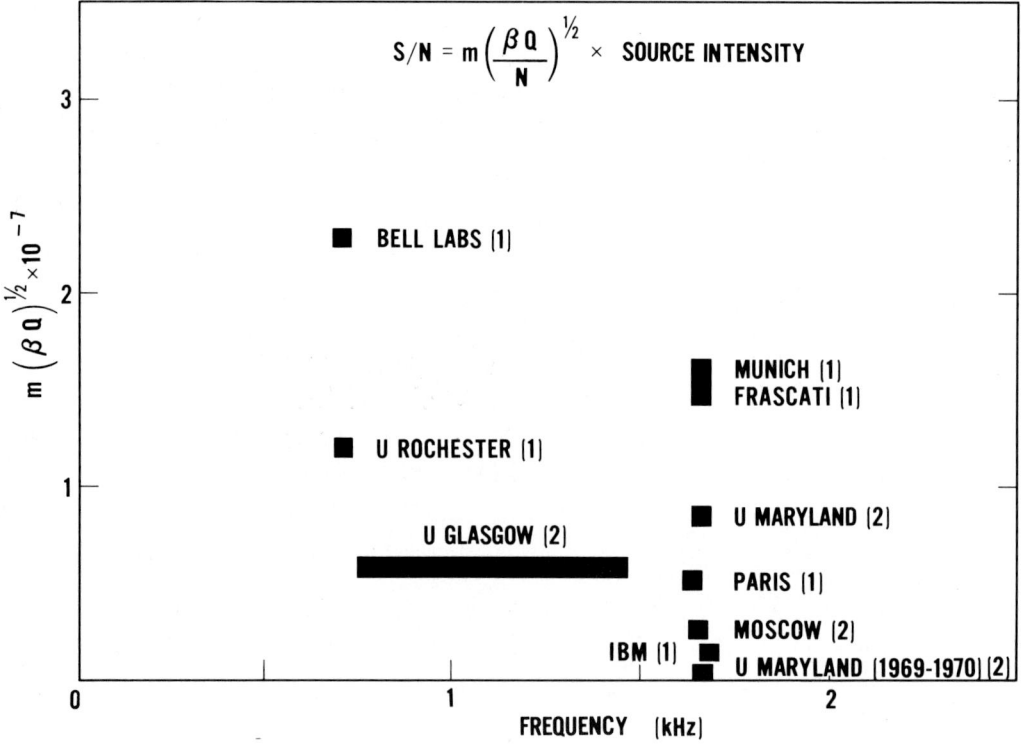

Fig. 8. Relative sensitivities to GR bursts of various antennas now operating, assuming equal amplifier noises and similar data analysis. Although the parameter $m(\beta Q)^{1/2}$ will not be easily changed, much better total sensitivity will be reached soon due to lower noise preamplifiers.

Various searches have been made for other types of radiation coincident with the reported Weber events, all with null results. (See Slusher and Tyson, 1973.) The limiting electro-magnetic flux per Weber event ranges between about 10^{-21} erg cm^{-2} s^{-1} Hz^{-1} for UHF, microwave, and infrared searches to 10^{-24} erg cm^{-2} s^{-1} Hz^{-1} for gamma rays. Some of these searches concentrated only on the galactic center region. In terms of total pulse energy, in erg cm^{-2} event^{-1}, the positive and null GR observations are between 10^4 and 10^6, whereas electron neutrino null limits extend between 10^2 and 10^{-7}, muon neutrino limits $<10^{-4}$, gamma ray $<10^{-5}$, infrared $<10^{-9}$, and radio $<10^{-12}$. These upper limits put a severe constraint on any model of the possible source of large bursts of GR.

We now wish to present some preliminary results for the Bell Labs-U. Rochester two antenna coincidence experiment. The two antenna system has been operating at the design sensitivity (approx. 3 times Weber's present sensitivity) now for one month. Data are collected by totally independent tape systems, and these tapes are later correlated. Two independent analyses have been completed on part of this data, one by Bell Labs in collaboration with R. W. Lee (Stanford University) and one by the Rochester group. Both indicate no significant excess number of coincidences at zero time lag, with the two thresholds which we tried. These data are automatically calibrated in energy sensitivity and time by the introduction of artificial excitations of both antennas two seconds apart in real time, may times per week. Figure 9 shows the high threshold result. The ten standard deviation peak at +2 s is the calibration peak, in this case at $1kT$. This result implies that there were not more than about 10 GR events imparting more than $\frac{1}{2}kT$ energy to these bars during this 16 days.

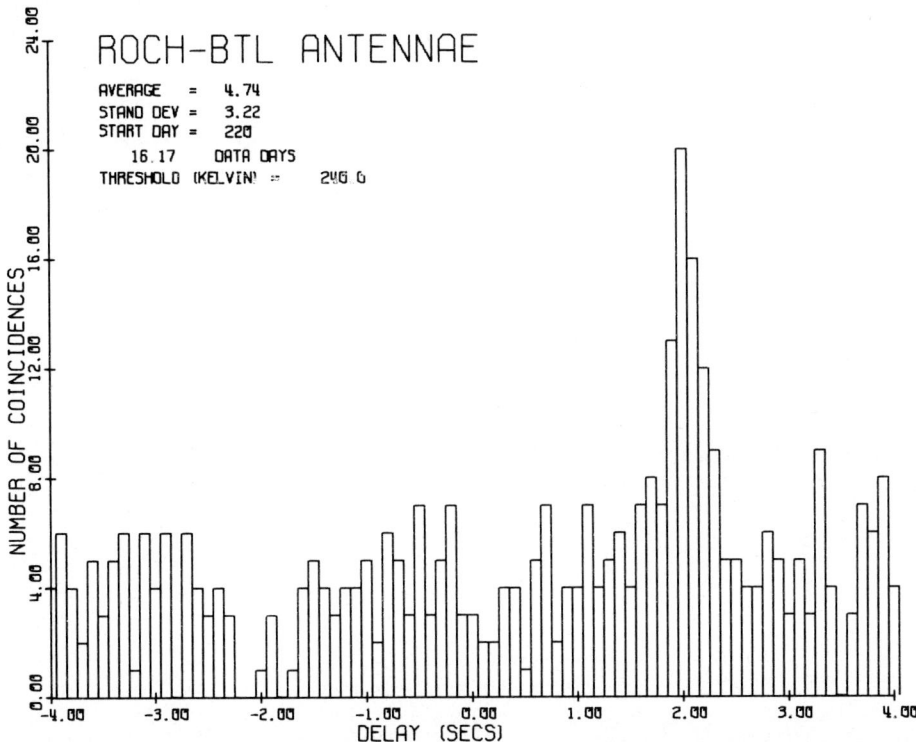

Fig. 9. Coincidence data for the Bell Labs-U. Rochester collaboration, using high threshold. No peak at zero lag is seen. The 10σ peak at +2 s lag is from 90 artificial excitations at a $1kT$ level introduced simultaneous plus 2 s.

Figure 10 shows the low threshold result in which 7006 and 4113 candidates were selected from each data tape by (as in Figure 9) convolving the raw amplitude and phase data with the time domain filter shown in Figure 11. This type of filter was also used during the last few years on our single antenna searches, and it has been very effective in eliminating sensitivity to occasional departures of amplitude or phase due to 'spikes' in the electronics. Returning to Figure 10, the 3σ peak at -3.8 s is not very significant, considering that there were 81 points in the lag plot with about 30 of them statistically independent. We are confident that there were few events above $\sim \frac{1}{4}kT$ ($\sim \frac{1}{10}kT$ in Weber's bars) during this time.

We measure the efficiency of our detectors at various levels directly. For a $1kT$ event, the Rochester antenna efficiency is 23% for ± 50 ms resolution and $\sim 90\%$ for

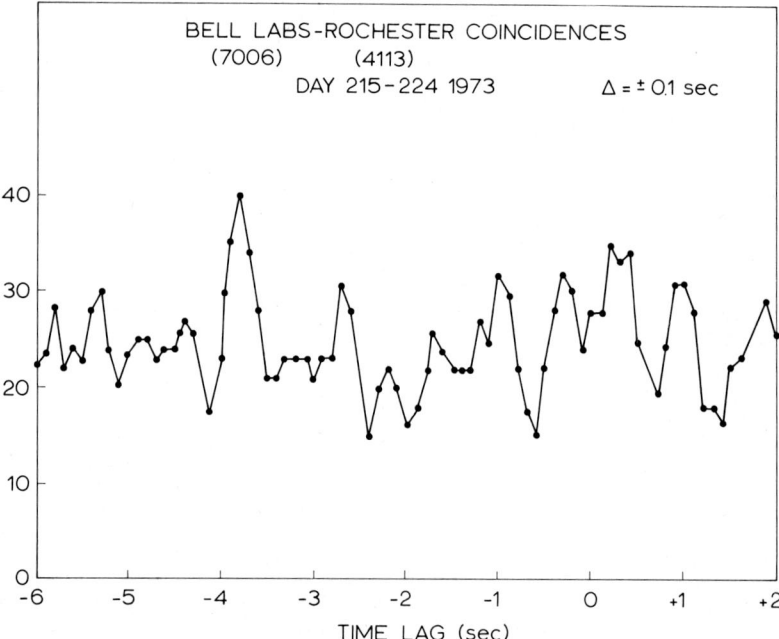

Fig. 10. Coincidence data for the Bell Labs-U. Rochester collaboration, using low threshold. No peak at zero lag is seen. No artificial calibration signals were applied.

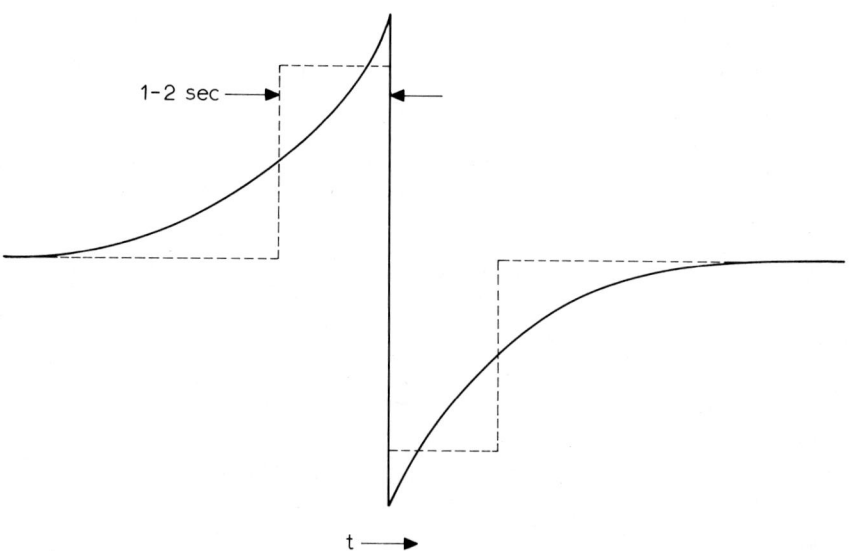

Fig. 11. Time-domain filter function used on magnetic tape data for the antenna coordinates X and Y. This filter is nearly optimal for the signature in Figure 6, and it discriminates against any electrical spikes.

±0.3 s resolution. The Bell Labs antenna is better. The efficiency of the two-antenna coincidence pair is ~85% at $1kT$ for ±0.3 s resolution. At event intensities of $\frac{1}{4}kT$, the Rochester efficiency is 15% and the Bell Labs efficiency 33%, for ±0.3 s resolution At $\frac{1}{70}kT$, the Bell Labs efficiency is $\gtrsim 1\%$. Or course, for GR, these antennas absorb ~3 times more energy from the wave, than Weber's antennas.

We emphasize that this experiment and Weber's experiment are sensitive to different signatures in the antenna amplitude or phase as a function of time; our analysis is sensitive to the kind of signature shown at the top of Figure 6, whereas Weber's current analysis is much more sensitive to 'delta functions' in the phase or amplitude as a function of time. The true origin of Weber's pulses remain an intriguing experimental problem worthy of our continued international effort.

The future holds possibilities for great improvements in sensitivity. We hope to push βQ for our antennas from the present value of 40 up to 10^4. Since the present calibrated noise temperature of our two antenna array is 20 K, this increase in sensitivity would give noise temperatures less than 1 K. We estimate that Weber's present antenna noise temperature is ~200 K, using his nonlinear algorithm. This implies that our antennas are ~10 times more sensitive to GR than Weber's present system, an increase of >100 over his system in 1970. But sensitivity increases of $>10^3$ may become possible with kilometer versions of the Hughes laser antenna (in space) or low-temperature designs. The suggestion of Burke (1973) may offer a unique output for a GR input. Prof. Braginsky will now suggest other possibilities.

References

Burke, W.: 1973, to be published in *Phys. Rev.*
Slusher, R. E. and Tyson, J. A.: 1973, *Nature* **243**, 25.
Tyson, J. A.: 1973, *Phys. Rev. Letters* **31**, 326.
Weber, J.: 1969, *Phys. Rev. Letters* **22**, 1320.
Weber, J.: 1970a, *Phys. Rev. Letters* **24**, 167.
Weber, J.: 1970b, *Phys. Rev. Letters* **25**, 180.
Weber, J.: 1972, *Nature* **240**, 28.

THE PROSPECTS FOR HIGH SENSITIVITY GRAVITATIONAL ANTENNAE

V. B. BRAGINSKY

Dept. of Physics, Moscow State University, Moscow, U.S.S.R.

Abstract. The sensitivity of a resonant gravitational wave detector which is necessary for the detection of pulses from asymmetric stellar collapses or from black-hole collisions in nearby galaxies is examined. Limitations on this sensitivity due to the resonating quality of the detector, thermal noise, and reaction of the detection system upon the detector are studied. No severe difficulties for the detection of the above-mentioned pulses are expected. For the detection of pulses from a cluster at the galactic center, a nonresonant system based on Doppler ranging to a drag-free satellite would be more appropriate. At present the sensitivity of Doppler-ranging is still two orders of magnitude below the requirements.

(I) The collision of two collapsed stars ($M \simeq 10\ M_\odot$) or of two neutron stars, and the nonsymmetric collapse of one star ($M \simeq 10\ M_\odot$) give a pulse of gravitational radiation with energy $E_g \simeq 10^{52}\text{--}10^{54}$ erg, duration $\hat{\tau}_g \simeq 10^{-3}\text{--}10^{-4}$ s and mean frequency $W_g \simeq 10^4\text{--}10^5$ rad s^{-1} (Zel'dovich and Novikov, 1964; Thorne, 1967; Burke and Thorne, 1969; Braginsky et al., 1969). To realize a reasonable experiment to detect such pulses it is necessary that an experimentalist observe at least 10 events (coincidences in two or more antennae) during one year. Taking account of the fact that the density of galaxies is approximately 3 per (Mpc)^3, one can obtain a simple connection between the sensitivity of the gravitational wave detector $\tilde{I}$ [erg cm^{-2}], the number of the galaxies N from which gravitational-wave burts can be observed and the frequency Ω of collisions and collapses in one galaxy collisions and collapses for which $E_g \simeq 10^{52}\text{--}10^{54}$ erg).

The Table I shows some value of $\tilde{I}$, N, Ω for three distances $R_1 = 3 \times 10^{25}$ cm, $R_2 = 3 \times 10^{26}$ cm, $R_3 = 3 \times 10^{27}$ cm.

TABLE I

10 events observed by experimentalist per year ($\varepsilon_g \simeq 10^{+52}\text{--}10^{+54}$ erg, $\hat{\tau}_g \simeq 10^{-3}\text{--}10^{-4}$ s)

R (distance)	N (the number of the galaxies in a sphere $\frac{4}{3}\pi R^3$)	Ω (the frequency of events in one galaxy)	$\tilde{I}$ (the density of grav. wave energy near the antenna)
3×10^{25} cm = 10 Mpc	$\simeq 10^{+4}$	$\simeq 10^{-3}$ yr^{-1}	10^{+1} erg cm^{-2}
3×10^{26} cm = 10^2 Mpc	$\simeq 10^{+7}$	$\simeq 10^{-6}$ yr^{-1}	10^{-1} erg cm^{-2}
3×10^{27} cm – optical horizon	$\simeq 10^{+10}$	$\simeq 10^{-9}$ yr^{-1}	10^{-3} erg cm^{-2}

Thus if a terrestrial gravitational antenna registers pulses of flux $\tilde{I} \simeq 10^{-3}$ erg cm^{-2}, and these pulses appear $\simeq 10$ time per year, then it is possible to conclude that probably collisions of black holes or neutron stars take place approximately one time in a billion years in one galaxy.

Conventional astrophysics suggests that a sensitivity of $\tilde{I} = 10^{+1}$–10^{-3} erg cm^{-2} for $\hat{\tau} \simeq 10^{-3}$ s, seems to be reasonable for the time scheme of coincidence of Weber (1969, 1970a, b). This required sensitivity is much better than the sensitivity which has been achieved in the recent experiments (Bonazolla, 1973; Tyson, 1972; Braginsky et al., 1970; Kafka, 1973). Let us examine the conditions necessary for obtaining such sensitivity.

The equation of motion for a quadrupole gravitational wave detector

$$m\ddot{X}^\mu + H\dot{X}^\mu + KX^\mu = -mc^2 l^\alpha R^\mu_{0\alpha 0} + F_N, \qquad (1)$$

where K is the rigidity which connects the two test masses, separated by a distince l, H is the coefficient of friction, $mc^2 l^\alpha R^\mu_{0\alpha 0}$ is the force produced by the gravitation wave, $R^\mu_{0\alpha 0}$ is the ac component of the Rieman tensor, c is the speed of light, F_N is the fluctuational force. In the case of pure thermal fluctuation the spectral density $(F_N)^2_f$ is equal to

$$(F_N)^2_f = 4\varkappa T_M H, \qquad (2)$$

where $\varkappa$ is the Boltzman constant, T_M is the temperature of the heat bath in which the mechanical part of the detector is situated. The ratio of the two forces on the right side of the Equation (1) is the ratio of signal to noise. Using the spectral density $\tilde{I}$ instead of $R^\mu_{0\alpha 0}$, assuming that the orientation of the detector relative to the wave's polarization and the propagation directions is optimal, and assuming that the signal to noise ratio is equal to one, we obtain an equation for the minimum detectable flux $[\tilde{I}]_{min}$ in one pulse:

$$[\tilde{I}]_{min} \simeq \frac{c^3 \varkappa T_M}{2\pi G m \omega_g Q_M l^2}. \qquad (3)$$

Here $\omega_g \simeq 1/\hat{\tau}_g$ if the mean frequency of gravitational radiation in the pulse, $\omega_g \simeq \omega_M = \sqrt{K/m}$, $Q_M = m\omega_M/H$ is the mechanical quality factor.

Equation (3) is also valid if $\hat{\tau} \ll \tau^*_M = 2Q_M/\omega_M$ (see details in Braginsky, 1970). It is important that the Equation (3) is obtained under the assumption that the registering system is ideal (about the backward fluctuational influence of the registering system see below). Equation (3) gives the *first important condition* for the detection of small gravitational wave pulses.

The simple conclusion that one may obtain from Equation (3) is the following: in order to increase the sensitivity it is necessary to minimize T_M and (or) to increase m and Q_M. Substituting into Equation (3) $m = 10^{+6}$ g, $m_g = 10^{+4}$ rad s^{-1}, $l = 10^2$ cm, $Q_M = 2 \times 10^5$, $T_M = 3 \times 10^{-3}$ K (the parameters of the installations for the experiments of Fairbank and Hamilton (1970)) we obtain $[\tilde{I}]_{min} \simeq 1$ erg cm^{-2}. This is apparently the limit of sensitivity which may be obtained with heavy aluminium bars, since the Q_M of aluminium even at low temperature, is not very high.

If we use a dielectrical monocrystal instead of aluminium, the Q_M may be much greater. It is possible to show that in a dislocation-free monocrystal without surface losses (the surface losses can be eliminated (by careful polish) the product $Q_M \omega_M$ is

$$Q_M \omega_M \simeq \frac{4 \varrho C_p^2}{\beta T_M \alpha^2}, \qquad (4)$$

where ϱ is the density; C_p, the heat capacity; α, the linear expansion coefficient; β, the heat conductivity.

The Table II includes the estimates for the factor $Q_M \omega_M$ of sapphire (Al_2O_3) at different temperatures based on the Equation (4) and experimental data for β, C_p and α (Yates et al., 1972; Goer and Dreyfus, 1967; Wolfmeyer and Dillinger, 1971). Table II also include estimates of $[\tilde{I}]_{min}$ for a relatively small sapphire mono-crystal ($m = 4 \times 10^4$ g, $l = 40$ cm), The estimates $[\tilde{I}]_{min}$ in Table II for an ideal monocrystal bar look very optimistic. Below we shall see that another type of difficulty will appear if an experimentalist has an ideal monocrystal at low temperature and wants to obtain sensitivity better than $[\tilde{I}]_{min} \simeq 10^{-2}$ erg cm^{-2}.

TABLE II

$T =$	300 K	80 K	4.2 K	2 K	0.4 K	0.01 K
$Q_M \omega_M$ rad s^{-1}	2.2×10^{16}	8.4×10^{15}	2.8×10^{17}	2×10^{18}	10^{21}	$\sim 3.5 \times 10^{22}$
$[\tilde{I}]_{min}$ erg cm^{-2}	5	3.5	5×10^{-3}	3.5×10^{-4}	1.5×10^{-7}	$\sim 10^{-10}$

Table III shows the preliminary results of measurement of for sapphire obtained by H. Bagdasarov, V. Mitrofanov and by the author. The sapphire monocrystal cylinder has $L = 15$ cm, $D = 4.5$ cm, $M = 1.1 \times 10^{+3}$ g. The results in Table III show that our monocrystal cylinder was not ideal (the factor $Q_M \omega_M$ at 80 K is two order smaller than the theoretical limit). But if we should decide to make a gravitational

TABLE III
$(\omega_M = 2 \times 10^5 /\text{rad s}^{-1}) (f_M = 34 \text{ kHz})$

T	300 K	80 K
Q_M	4×10^7	1.3×10^8
$Q_M \omega_M$	8×10^{12}	2.6×10^{13}
$\tau_M^* = \dfrac{2 Q_M}{\omega_M}$	400 s	1200 s

antenna with this small cylinder we may reach approximately the sensitivity of a heavy aluminium bar which has $Q_M \omega_M \simeq 2 \times 10^9$ (Weber 1969, 1970a, b; Bonazolla, 1973; Braginsky et al., 1972; Tyson, 1972; Kafka, 1973).

The *second condition* necessary for the detection gravitationalwave pulses is to construct a very sensitive system for registering small vibration of the detector. The ferroelectrical transducers seem not to be suitable for $\tilde{I}$ less than 10 erg cm^{-2}. Active transducers (modulator-demodulator type) are more promising. Study of different types of active transducers is now in progress in many groups (Braginsky et al., 1972; Dick and Yen, 1972; Paik, 1972).

Figure 1 shows the scheme of an active capacity choice of system connected with a quadrupole mechanical oscillator.

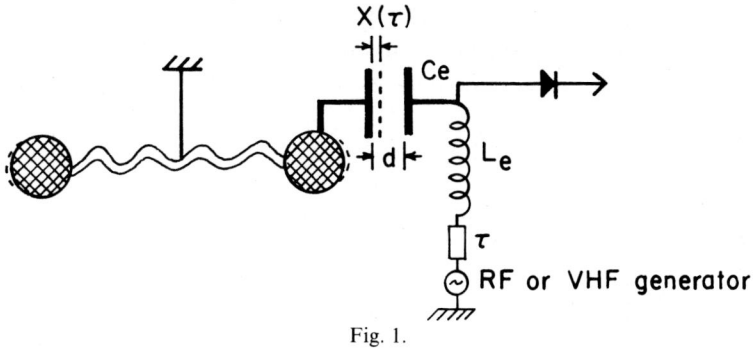

Fig. 1.

If it is possible to eliminate (using a compensation scheme) the noise of the RF generator, then the smallest displacement $[X(\tau)]_{min}$ between the plates of the capacity depends on the thermal fluctuation in the electrical resonator and on the amplitude $U_\sim$ of RF voltage on the plates (see details in Braginsky, 1970):

$$[X(\tau)]_{min} \simeq \frac{2d}{U_\sim} \sqrt{\frac{4\varkappa T_e \Delta f}{\omega_e Q_e C_e}}. \tag{5}$$

Here d is the mean distance between the plates; $\varkappa$, the Boltzman constant; T_e, the temperature of the electrical resonator; $\omega_e = 1/\sqrt{L_e C_e}$, the eigen frequency of the resonator; $Q_e = \omega_e L_e/\tau$, the quality factor of the resonator; C_e, the capacity; $\Delta f \simeq 1/\tau$. Substituting into Equation (5) the values $T_e = 2$ K, $W_e = 6 \times 10^{10}$ rad s^{-1}, $C_e = 10^{-12}$ F, $Q_e = 10^{+11}$ (this Q_e has been achieved by Weissman and Turneaure (1968) we obtain $[X(\tau)]_{min} \simeq 10^{-21} \sqrt{\Delta f}$ cm.

The pulse of gravitational radiation which lasts for a time $\hat{\tau}_g$ and has an energy flux of energy $\tilde{I}$ produces a change in the oscillation amplitude of the detector $X_g(\tau)$ equal to

$$X_g(\tau) \simeq l \sqrt{\frac{8\pi G}{C^3} \tilde{I} \hat{\tau}_g}, \tag{6}$$

if $\omega_M \simeq \omega_g$ and $\hat{\tau}_g \ll \tau_M^* = 2Q_M/\omega_M$.

Three estimates of $X_g(\tau)$ for three different $\tilde{I}$ are given in Table IV.

TABLE IV

($l = 40$ cm, $\hat{\tau}_g = 10^{-3}$ s)

$\tilde{I}$ [erg cm^{-2}]	10^{-3}	10^{-1}	10^{+1}
$X_g(\tau)$ [cm]	$\simeq 1 \times 10^{-20}$	$\simeq 1 \times 10^{-19}$	$\simeq 1 \times 10^{-18}$ cm

Comparing the data in Table IV with the estimate $X(\tau) \simeq 10^{-21}\sqrt{\Delta f}$ cm based on the present state of art for Q_e, one can reach a preliminary conclusion that it is possible to reach the sensitivity $\tilde{I} = 10^{-3}$ erg cm^{-2} for millisecond gravitational wave pulses. But a *third condition* for a detector exists. In Equations (3), (5) and (6) the reaction of the registering system on the masses of the detector is not taken into account (see details in Braginsky, 1970; Braginsky, 1973). It is possible to show that if $Q_M = \infty$, definite classical limits exist for the minimum detectable force $[F(\tau)]_{\min}$ acting on a mechanical oscillator. These limits depend on the relations between $\hat{\tau}$, $\tau_e^* = 2Q_e/\omega_e$ and ω_M.

If $\hat{\tau} \gg \tau_e^*$, and $1/\omega_M \gg \tau_e^*$, then

$$[F(\tau)]_{\min} \simeq \frac{4}{\hat{\tau}} \sqrt{m \varkappa T_e \frac{\omega_M}{\omega_e}}, \tag{7}$$

and if $\hat{\tau} \ll \tau_e^*$, and $\tau_e^* \gg 1/\omega_M$, then

$$[F(\tau)]_{\min} \simeq \frac{4}{\hat{\tau}} \sqrt{m \varkappa T_e \frac{\omega_M}{\omega_e} \frac{2\hat{\tau}}{\tau_e^*}} = 4\sqrt{\frac{m \varkappa T_e \omega_M}{\hat{\tau} Q_e}}. \tag{8}$$

The Equations (7) and (8) are classical, they are valid if Q_M is high enough (when the estimates from (7) or (8) are greater than estimates from Equation (2)). Equations (7) and (8) show that for to minimize $[F(\tau)]_{\min}$, it is necessary to decrease the ratios ω_M/ω_e and $\hat{\tau}/\tau_e^*$, and to increase the Q-factor $Q_e = \tau_e^* \omega_e/2$ of the electrical resonator.

The estimates for $[F(\tau)]_{\min}$ from Equation (8) are less than those from Equation (7), if $\hat{\tau} \ll \tau_e^*$. Equation (8) will not be valid if $[F(\tau)]_{\min}$ from (8) is equal or less than $[F(\tau)]$ quantum:

$$[F(\tau)]_{\text{quantum}} \simeq \frac{4}{\hat{\tau}} \sqrt{\frac{m \hbar \omega_M}{n}}, \tag{9}$$

where $\hbar$ is the Planck constant, n is the quantum level of the mechanical oscillator before $F(\tau)$ acted on it. Equation (9) is based on the uncertainty principle (see details in Braginsky (1970, 1973)). Assuming $[F(\hat{\tau}_g)]_{\min}$ from Equations (7) and (8) to be equal to $mc^2 l^\alpha R_{0\alpha 0}^\mu$, it is easy to obtain two new limits for $\tilde{I}$ which are valid for high Q_M:

$$[\tilde{I}]_{\min} \simeq \frac{2c^3 \varkappa T_e}{\pi G m \omega_M \omega_e l^2 \hat{\tau}_g}, \tag{10}$$

if $\hat{\tau}_g \gg \tau_e^*$ and $1/\omega_M \gg \tau_e^*$;

$$[\tilde{I}]_{\min} \simeq \frac{c^3 \varkappa T_e}{\pi G m \omega_M Q_e l^2}, \tag{11}$$

if $\hat{\tau}_g \ll \tau_e^*$ and $1/\omega_M \ll \tau_e^*$.

Equation (3) is similar to Equation (11), but in (11) the ratio T_M/Q_M is replaced by the ratio T_e/Q_e.

Substituting into Equation (11) the values $m = 4 \times 10^4$ g, $l = 40$ cm, $\omega_M = 10^{+4}$ rad s^{-1}, $T_e = 0.01$ K and $Q_e = 10^{+12}$, we obtain $[\tilde{I}]_{min} \simeq 10^{-2}$ erg cm^{-2}. This estimate is very promising, but at the same time it is much worse than the estimate 10^{-10} erg cm^{-2} for the same m, W_M and T_M (see Table II) which is based on Equation (3). In other words the third condition which gives Equation (10) and (11) is more severe, and for to reach a sensitivity better than $\tilde{I} \simeq 10^{-2}$ erg cm^{-2}, it is necessary to have $T_e < 10^{-2}$ K, and (or) $Q_e > 10^{+12}$.

Equations (3), (7), (8), (10) and (11) were obtained under the assumption that the time of measurement (the resolution time) τ_{res} is equal to the expected duration of the gravitation wave pulse τ_g. If one chooses τ_{res} greater than τ_g it is easy to obtain new equations for $[\tilde{I}]_{min}$ which are similar to Equations (3), (10) and (11). These estimates for the sensitivity from these new equations are better than the estimates from Equations (3), (10) and (11). But in the case $\hat{\tau}_g \ll \hat{\tau}_{res}$ the experiment is not informative enough.

Concluding this review of various necessary conditions for achieving high-sensitivity detectors for the frequency $\omega_g \simeq 10^{+4} - 10^{+5}$ rad s^{-1} it is possible to say that the present state of experimental art and the present level of detector theory do not predict severe obstacles to observing the gravitational waves from relatively rare collisions of black holes and nonsymmetric collapsing stars.

(II) The programme and conditions described above are suitable for a definite level of Ω – the frequency of black holes collisions or nonsymmetric collapse. If the real frequency is less, and therefore the experimentalists cannot observe enough events per year to convince one that they are seeing anything we shall never obtain information about gravitational radiation from nonsymmetric collapsing systems.

Another probable source of gravitational radiation exists. Zel'dovich and Polnarev (1973) showed that, if clusters exist in the center of our Galaxy, they may produce pulses of gravitational radiation with intensity $\tilde{I} \simeq 10^{-3} - 1$ erg cm^{-2}, mean frequency $\omega_g \simeq 10^{-1} - 10$ rad s^{-1}, and duration $\tau_g \simeq 1$ s. This type of pulses may occur approximately 25 times in a year.

It is evident that the type of detector described above is not suitable for the frequency range $\omega_g \simeq 10^{-1} - 10$ rad s^{-1}. More fruitful perhaps is the idea (Braginsky and Hertzenstein, 1967) to use two free masses: the Earth and a drag-free satellite or two drag-free satellites with a Doppler ranging system between them.

The variation of the speed Δv_g of these two masses due to a gravitational wave pulse is equal to

$$\Delta v_g \simeq l \sqrt{\frac{8\pi G}{c^3} \frac{\tilde{I}}{\hat{\tau}_g}}, \qquad (12)$$

where l is the distance between the masses.

Equation (12) is valid if $l \lesssim \pi c / \omega_g$. It is easy to show that the thermal noise condition similar to Equation (3) is not important for this type of antenna, and it is necessary only to construct a Doppler ranging system. Substituting in Equation (12) the values $\tilde{I}=1$ erg cm^{-2}, $\hat{\tau}_g=1$ s, $l=10^{12}$ cm, we obtain $\Delta v_g \simeq 2.5 \times 10^{-7}$ cm s^{-1}. The present level of sensitivity (Anderson, 1972) is $\Delta v \simeq 10^{-5}$ cm s^{-1}. Thus, if experimentalists can improve the sensitivity of Doppler ranging systems about two orders, then it will be possible to realize this type of gravitational wave antennae.

References

Anderson, J.: 1972, Lecture at 'E. Fermi' Summer School of Physics, Varenna.
Bonazolla, S.: 1973, Lecture at the symposium, *L'ondes gravitationelles*, Paris.
Braginsky, V. B.: 1970, NASA Techn. Transl. TT-F 672, Physical Experiments with Test Bodies', 'Nauka'.
Braginsky, V. B. and Hertzenstein, M. E.: 1967, *Prisma JETP* **5**, 348.
Braginsky, V. B., Manukin, A. B., Popov, E. I., and Horev, A. A.: 1972, *JETP* **16**, 157; *Phys. Letters*, to be published.
Braginsky, V. B. and Vorontzov, Yu. I.: 1973, Preprint Inst. Theor. Phys., Kiev, to be published.
Braginsky, V. B., Zeldovich, Ya. B., and Rudenko, V. N.: 1969, *Pisma JETP* **10**, 437.
Burke, W. L. and Thorne, K. S.: 1969, OAP-184, 'Gravitational Radiation Damping, Preprint.
Dick, G. and Yen, H.: 1972, Report at the S. C. Conf. Annapolis, U.S.A.
Fairbank, W. and Hamilton, W.: 1970, *Proc. Conf. on Experim. Tests of Gravit. Theory, Caltech.* **38**, 85.
Goer, A. M. and Dreyfus, B.: 1967, *Phisica Status Solidi* **22**, 77.
Kafka, P.: 1973, Lecture at the symposium *L'ondes gravitationelles*, Paris.
Paik, H. Jung: 1972, Lecture at 'E. Fermi' Summer School of Physics, Varenna.
Thorne, K. S.: 1967, OAP-167, 'Non-Relativistic Pulsation of General-Relativistic Stellar Models, Preprint.
Tyson, J.: 1972, 'Null Search for Bursts of Gravitational Radiation', preprint.
Weber, J.: 1969, *Phys. Rev. Letters* **22**, 1320.
Weber, J.: 1970a, *Phys. Rev. Letters* **24**, 276.
Weber, J.: 1970b, *Phys. Rev. Letters* **25**, 180.
Weissman, I. and Turneaure, J.: 1968, *J. Appl. Phys.* **39**, 4417.
Wolfmeyer, M. and Dillinger, J.: 1971, *Phys. Letters* **34A** (4), 247.
Yates, B., Cooper, R. E., and Pojur, A. F.: 1972, *J. Phys. C. Solid State Phys.* **5**, 1046.
Zel'dovich, Ya. B. and Novikov, I. D.: 1964, *Dokl. Akad. Nauk* **155**, 1033.
Zel'dovich, Ya. B. and Polnarev, Yu.: 1973, 'The Radiation of Gravitational Wave from Clusters of High Density Stars, preprint.

GRAVITATIONAL RADIATION DETECTOR MAGNETIC TAPES FROM ROCHESTER AND MARYLAND*

M. LEE and J. WEBER

University of Maryland, College Park, Md., U.S.A.

Abstract. The Maryland Univac 1108 computer has analyzed magnetic tapes, one set having outputs of a 1661 Hz gravitational radiation detector at the University of Maryland and another set having outputs of a 710 Hz gravitational radiation detector at the University of Rochester. For a seven day period beginning August 23, 1973, 391 coincidences were found. Time delay experiments performed by the computer observed 338 accidental coincides. The zero delay excess is therefore 53 coincidences.

These data have not been verified by the Rochester group and any errors in this report are solely the responsibility of the Maryland authors.

---

* Supported in part by the U.S. National Science Foundation, and the University of Maryland Computer Science Center.

AN UPPER LIMIT TO THE MASS LOSS FROM THE CENTRE OF THE GALAXY*

ARCADIO POVEDA and CHRISTINE ALLEN

Instituto de Astronomía, Universidad Nacional Autónoma de México, México

Abstract. A mass loss of 200 $M_\odot$ per year, as conservatively suggested if Weber is detecting gravitational waves from an isotropic source at the galactic centre, is shown to be incompatible with the existence of (a) globular clusters, (b) old wide binaries, if this loss rate has been constant over the past 10^{10} yr.

From the orbit of ω Centauri in the galactic field and its observed mass distribution and tidal radius an upper limit to the mass loss from the galactic centre is found to be 1 $M_\odot$ yr^{-1} over the past 10^{10} yr.

* To be submitted to *Astrophys. J.*

OBSERVATIONS WITH WIDE-BAND GRAVITATIONAL RADIATION DETECTORS*

R. W. P. DREVER, J. HOUGH, R. BLAND, and G. W. LESSNOFF

Dept. of Natural Philosophy, University of Glasgow, Scotland, U.K.

Abstract. The principles and operation of wide-band gravitational radiation detectors are described. In 7 months of coincidence observations, one signal which fulfills requirements set for a gravitational wave event was recorded, and details are presented. The experiment sets an upper limit to millisecond pulses of gravitational radiation of 0.9 ± 2.1 per month at a 25% threshold of $0.3kT$ in 300 kG bars, which appears inconsistent with the flux implied by Weber's 1970 results if these are due to such pulses.

---

* This work has been submitted for publication in *Nature*. Part of it will be published in *Proceedings of the Symposium on Gravitational Waves and Radiation*, Paris, June 18–22, 1973. (To appear as C.N.R.S. Report No. 220.)

ON THE EVALUATION OF THE MUNICH-FRASCATI WEBER-TYPE EXPERIMENT*

PETER KAFKA

Max Planck Institut für Physik und Astrophysik, Munich, F.R.G.

Abstract. Both the theory of optimal evaluation, developed in Munich, and tests with artificial pulses suggest that the Munich-Frascati experiment was more sensitive to short gravitational pulses than the Maryland-Argonne experiment. This was the case even in April 1973, before the Munich sensitivity was increased by another factor of 5. Nevertheless, the coincidences between Munich and Frascati did not show any surplus number for zero time delay.

It is concluded that Weber's events very likely were not due to gravitational radiation. However, a tape with four days of data from Maryland and Argonne (June 1973), which Weber kindly sent to us, does show a peak at approximately zero time delay, about 0.3 s wide and centered at 0.1 s time delay. The significance is not quite 2 standard deviations for the zero delay bin, but higher when the bins are chosen 0.3 s wide. No explanation can be offered at the moment.

---

* A paper with the same title is contained in the proceedings of the Colloque International C.N.R.S. No. 220 *Ondes et radiations gravitationnelles*, Paris, 18–22 Juin 1973.

MEUDON GRAVITATIONAL RADIATION DETECTION EXPERIMENT

S. BONAZZOLA, M. CHEVRETON and J. THIERRY-MIEG

Observatoire de Paris, Meudon, France.

Abstract. A description of Weber's type experiment of gravitational waves detection is given. Double and triple coincidences with Frascati and Munich groups were looked for with negative results. A capacitive passive method of detection has been presented. This method can improve the sensitivity by a factor of 30 when working at room temperature.

THE USE OF CRYOGENIC TECHNIQUES TO ACHIEVE HIGH SENSITIVITY IN GRAVITATIONAL WAVE DETECTORS

S. P. BOUGHN, W. M. FAIRBANK, M. S. McASHAN, H. J. PAIK, and R. C. TABER

Dept. of Physics, Stanford University, Stanford, Calif. 94305, U.S.A.

and

T. P. BERNAT, D. G. BLAIR, and W. O. HAMILTON

Dept. of Physics and Astronomy, Louisiana State University, Baton Rouge, La. 70803, U.S.A.

Abstract. Cryogenic detectors for gravitational wave astronomy promise greatly improved sensitivity over room temperature detectors. The 3 mK detector which we have under construction should give an improvement of 10^6 over existing detectors. The cryogenic antennae are described and the calculated low temperature performance is detailed. New superconducting instrumentation is described.

1. Introduction

We are embarked on a cooperative program between Stanford University, Louisiana State University and the University of Rome to build large scale gravity wave detectors at each location which can be cooled to the lowest attainable temperature, hopefully the order of 3×10^{-3} K, 3 mK. The calculated sensitivity of these detectors for gravity waves is 0.1 erg cm^{-2} Hz^{-1} (Fairbank *et al.*, 1972; Paik, 1972), which would represent an improvement of 10^6 over existing room temperature detectors. This level of sensitivity would make possible the detection of gravitational wave signals at the level predicted from known astronomical sources, and would offer improved resolution on the events reported by Weber.

In 1959 Joseph Weber (Weber, 1960, 1961) started an experimental program with the purpose of detecting gravitational radiation. His pioneering work during the 1960's led to the development of a mass quadrupole antenna capable of detecting gravitational wave signals with an energy as small as 10^5 ergs cm^{-2} Hz^{-1}. Since 1969 Weber has reported evidence that he is receiving signals of roughly this order of magnitude (Weber, 1970). The large energy associated with these signals have prompted theorists to again look into possible sources of gravitational radiation. We are hearing at this meeting about the results from other room temperature detectors built with similar sensitivities to Weber's.

The most likely theoretical source of high intensity gravitational radiation is gravitational collapse to a neutron star or black hole (Hewish *et al.*, 1968), Supernova, which are thought to be the result of collapse of a star to a neutron star are observed in other galaxies at the rate of one supernova every 30 yr. This is one every few days in the Virgo cluster of 2500 galaxies, but hopelessly rare in a single galaxy. Recently the discovery of X-ray sources (Giaconni *et al.*, 1972) and their binary nature has given us a tool not only to measure the masses of neutron stars but also to identify other collapsed objects, namely black holes (Ruffini, 1972). It has been postulated that many collapses can take place in binary stars without supernovae. The estimate of collapsed objects which are neutron stars or black holes has been put at approximately 10^9 in our own galaxy on the basis of the experimental evidence. Recently the discovery of bursts of γ rays from outside our solar system (Klebesadal et al., 1973) with energies of 10^{-3} ergs cm^{-2} per burst at the surface of the Earth has raised the question of whether these could be associated with gravitational collapse in our Galaxy. Important theoretical research has paralleled the experimental breakthroughs involving gravitational collapse. The theory predicts in all cases the emission of an energy between 10^{49} and 10^{52} ergs per burst (Ruffini, 1972).

The calculated increase of 10^6 over existing room temperature detectors would enable one to see the details of the bursts of gravitational radiation emitted by collapsing objects in our own Galaxy. This would also make possible the detection of strong signals of the order of a solar mass, 10^{54} ergs, from the nearby 1000 galaxies. It is important to emphasize that there is no better way to inquire about the formation of a collapsed object than to look for the millisecond gravitational waves emitted in the process of formation. This also provides a possible tool to check the relativistic analysis of radiation from particles falling into black holes.

In this paper we wish to show the reasons why a low temperature detector can be built with a theoretical improvement of 10^6 over existing room temperature detectors.

The program to build such detectors at Stanford, Louisiana State University and Rome envisions two and eventually three detectors at each location. This would allow one to monitor continuously all of the observable sources over the sky 24 h a day. Interferometric measurements would give clear indications of the direction of the source for signals sufficiently above noise to determine the phase of oscillation of the individual antennas. The achievement of increased sensitivity might make possible a correlation between gravitational events and astronomical events in the electromagnetic spectrum, providing a basis for a new gravitational wave astronomy.

2. Sensitivity of a Cryogenic Detector

In order to understand the source of the great improvement in the sensitivity of gravitational wave detectors made possible by cooling both the antenna and the receiver to very low temperatures, let us begin with a review of the expressions for the sensitivity of existing Weber type room temperature detectors using piezoelectric crystals as transducers to transform the mechanical energy in the aluminum gravity wave antenna to an electrical signal in the amplifier.

The energy absorbed by the nth vibrational mode of a noiseless gravitational wave antenna of cylindrical shape from an incoming gravitational wave pulse of energy density $F(v_n)$ (erg cm^{-2} Hz^{-1}) emitted by a favorably polarized source whose line of sight is perpendicular to the longitudinal axis of the antenna can be shown to be (Ruffini and Wheeler, 1971)

$$E_s = \left[\frac{8}{\pi} \frac{1}{n^2} \frac{G^2}{c} \frac{v_s^2}{c^2} \right] F(v_n) M, \qquad (1)$$

where M is the total mass of the antenna, v_s is the speed of sound in the antenna. G is the gravitational constant and c is the velocity of light. n will be set equal to 1 in the following discussion. In this case, for aluminum the above formula becomes

$$E_s = 1.6 \times 10^{-27} F(v_n) M \text{ ergs cm}^{-2} \text{ Hz}^{-1}. \qquad (2)$$

To determine the minimum detectable signal one is interested in the ratio E_s/E_N where E_N is the effective noise energy in the received signal. For Weber type detectors with a piezoelectric pickup it has been shown that (Gibbons and Hawking, 1971; Tyson, 1971; Paik, 1972)

$$\frac{E_s}{E_N} = 1.6 \times 10^{-27} F(v_n) M \left[\frac{\frac{\pi}{2} \beta Q_A Q_R}{k^1 T_A T_R} \right]^{1/2}, \qquad (3)$$

where Q_A is the antenna Q, β is the coupling coefficient of the antenna to the transducer, T_A is the noise temperature of the antenna, T_R is the noise temperature of the receiver, k is Boltzmann's constant and Q_R is the loaded Q of the receiver adjusted for optimum E_s/E_N.

For a piezoelectric crystal of the type used by Weber,

$$\frac{1}{Q_R} = \tan \delta + \omega_n R_s C \qquad (4)$$

where $\tan \delta$ is the dissipation factor of the transducer, ω_n is the angular resonant frequency of the antenna, R_s is the series noise resistance of the preamplifier and C is the capacitance of the piezoelectric transducer. Equation (3) suggests all the essential features of the optimization procedure for a Weber type detector. To increase the signal strength, one should increase M consistent with keeping the fundamental resonance in the millisecond region and maximize the term in the brackets.

By cooling to very low temperatures and using a superconducting resonant diaphragm in place of the quartz crystal and a superconducting SQUID (Superconducting Quantum Interference Device) magnetometer with a measured noise temperature less than 10^{-4} K as a parametric amplifier, it is possible to improve not only T_A but also T_R and $\beta Q_A Q_R$. Paik (1972) has analyzed in detail the superconducting diaphragm and SQUID magnetometer being built at Stanford for the 3 mK gravitational wave detector and shown that $\beta Q_A Q_R / T_A T_R$ is theoretically improved over the parameters reported by Weber (1972) for a room temperature antenna by 10^{11}. This coupled with an increase in M from 1.4×10^3 kg to 5×10^3 gives an overall improvement in E_s/E_N of 10^6. This gives for E_s a minimum detectable signal 0.1 erg cm^{-2} Hz^{-1}. An untuned version of the superconducting detector is being built at Louisiana State University. The calculated sensitivity of this detector is equivalent. A third version of the detector is being designed and built at Rome.

An E_s of 0.1 erg cm^{-2} Hz^{-1}, as predicted for the 3 mK gravitational wave detector represents a flux which would be produced by a source at the center of our galaxy which converts $7 \times 10^{-7} M_\odot$ or 1.2×10^{48} erg into gravitational radiation in 1 ms

(1 kHz bandwidth). Therefore, as was discussed in Section 1 this detector will enable us to analyze the pulse shape and the polarization of the predicted signals coming from our Galaxy. It should also make possible the determination of the direction of the source and the velocity of propagation of gravity waves through the use of gravitational wave interferometry.

3. Superconducting Transducer

3.1. Resonant Diaphragm with SQUID

The principle of the tuned cryogenic detector is as follows. The energy of the antenna in the observed mode is coupled out to a small resonant mass (niobium diaphragm) tuned to the antenna resonant frequency. The motion of this resonant mass modulates the inductance of a superconducting loop carrying a dc current and this produces an ac magnetic field which is in turn detected with a low-noise superconducting magnetometer. For the details of this system, see the paper by Paik (1972). A schematic diagram of the transducer along with a symbolic representation of the magnetometer is shown in Figure 1. Two flat coils L_1 and L_2 are wound with niobium-titanium wire 0.005 cm in diameter and 0.01 cm away from the two superconducting surfaces

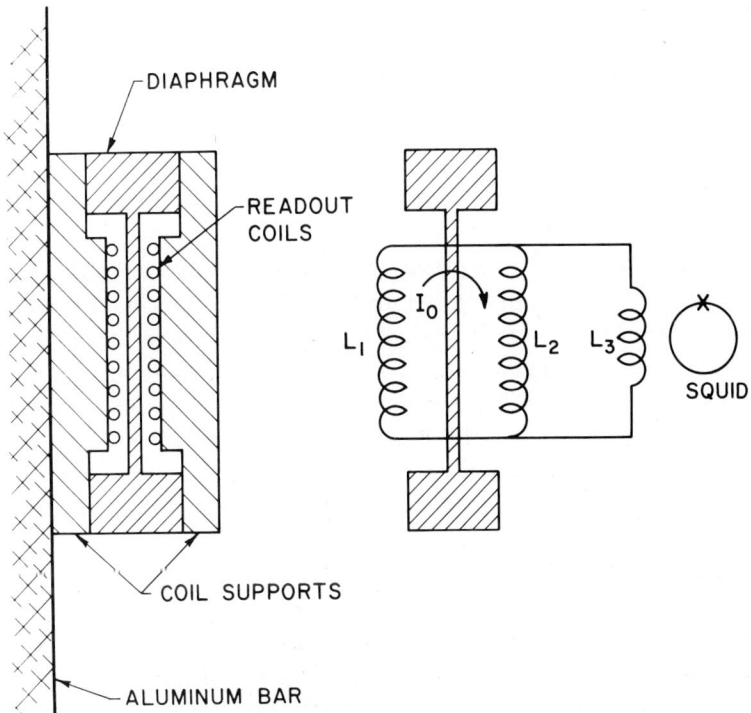

Fig. 1. Super conducting resonant transducer and SQUID magnetometer.

of the niobium diaphragm. These coils and the circular edge of the diaphragm are rigidly clamped to the end of the aluminum bar. A large persistent current of the order of 5 A is stored in the superconducting loop formed by L_1 and L_2. A third coil L_3 also wound with niobium-titanium wire is connected in parallel with L_1 and L_2 and fed to the SQUID which serves as a parametric amplifier.

As the diaphragm moves back and forth, the inductances L_1 and L_2 are modulated and as a result of flux quantization in the two superconducting loops a net ac current appears in L_3. In turn the magnetic field produced by this current in L_3 is detected by the SQUID magnetometer. The dc current stored in L_1 and L_2 plays the following important roles.

(1) It tends to push the diaphragm away from the two coils L_1 and L_2 thus centering the equilibrium position and allowing the desirable spacing 0.01 cm on both sides of the diaphragm.

(2) The ac energy that appears in the circuit as a result of the diaphragm motion acts as an additional spring by which one can tune the resonant frequency of the diaphragm to the antenna frequency.

(3) A high coupling γ is obtained by means of a large dc current. A persistent current of the order of 5 A should produce $\gamma = 0.95$ and increase the resonant frequency of the diaphragm from 300 Hz to 1350 Hz. We have experimentally confirmed this dependence of the energy coupling and the resonant frequency on the stored dc current.

As is shown by Paik (1972), a high mechanical and electrical Q of the resonant diaphragm is essential in order not to degrade the high Q of the antenna. We have obtained a mechanical Q as high as 3×10^6 after electropolishing the niobium surface and heat-treating it for 10 h at 1800 °C. No dependence of the Q of the transducer on the stored dc current was observed in the low coupling region of $\gamma \lesssim 0.03$. The Q of the transducer remained high when coupled to the SQUID magnetometer amplifier.

It is possible to mount on each gravity wave antenna three resonant diaphragms tuned to the frequencies of the first three longitudinal modes as is indicated in Figure 5. The outputs from the three transducers can be fed into three separate magnetometers. Because of its quadrupole character, a gravitational wave will excite only odd harmonics whereas seismic and most other disturbances will excite all the modes within the frequency spectrum of the pulse. Equation (1) shows that the cross section for the nth mode is reduced by $1/n^2$ from the value for the fundamental mode. So our detector should still be sensitive to these higher modes and give additional information on the pulse shape of a signal.

The magnetometer sensor (SQUID) to be used in our tuned detectors is a toroidal cavity cut between two solid niobium blocks with a weak link at the center. The noise temperature of a SQUID magnetometer has been measured to be less than 1 mK (see Zimmerman, 1972; Giffard et al., 1972). For details of the principles of operation and present state of the art of 30 MHz SQUID magnetometers see the above article and the article by Silver and Zimmerman (1967).

3.2 Single axis superconducting accelerometer

We have also constructed a uniaxial accelerometer which has sufficient sensitivity to detect the Brownian motion of the detector bar at 2 K and which should, when the temperature is reduced, detect the bar motion at 3 mK. The detector diagram is shown in Figure 2.

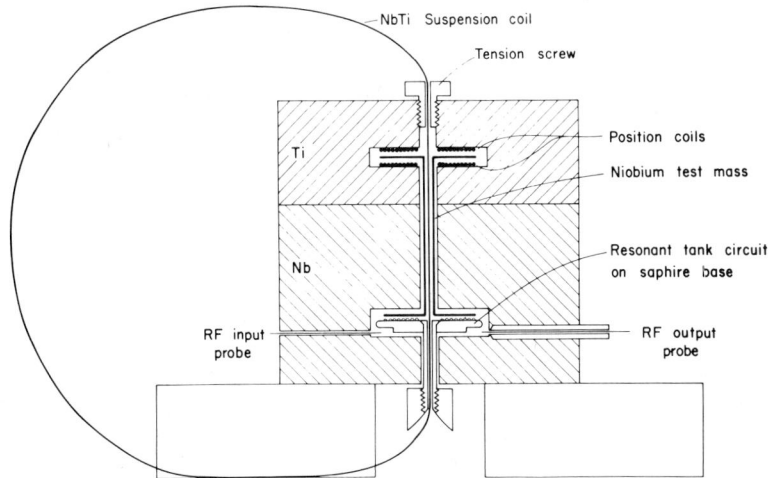

Fig. 2. Non-resonant superconducting transducer.

A superconducting wire carries a persistent current whose magnetic field acts to levitate the niobium spool. The spool then is tightly bound in directions perpendicular to the wire and can move without friction along the wire. If the accelerometer case is bounded to the end of the bar with the wire perpendicular to the bar, the case will move with the bar while the spool maintains a fixed position in inertial space. A superconducting inductance is mounted on the case and its inductance is then modulated due to relative motion of the case and the niobium spool. The position of the spool with respect to the inductance can be adjusted by the position coils at the back of the spool.

The sensitivity of this accelerometer is determined by the accuracy to which a small change of inductance can be measured. The inductance can be made a part of a resonant circuit so that the resonant frequency changes with the motion of the bar. The use of superconducting circuits allows this sensitivity to be very high because of the high Q which can be obtained for superconducting circuits.

We have constructed a prototype model of this accelerometer and obtained a Q in excess of 10^4 for the varied inductance which enables us to measure bar deflections as small as 10^{-15} cm at 2 K. This calculated sensitivity has not yet been directly measured because the lack of vibration mounting on the prototype caused an excessive mechanical noise level. This situation is being improved and new measurements are underway. We have also constructed a low noise MOSFET amplifier which works while immersed in liquid helium and are using this to detect the amplitude change of the tuned circuit when it is driven by a constant current generator. Detailed calculations of the performance of this detector when mounted on the 3 mK gravitational antenna indicate a sensitivity 30 dB below kT providing that the oscillator noise can be effectively reduced or balanced out. This sensitivity is comparable to that of the resonant detector already described. The position coils can also be actuated so as to resonate the spool at the mechanical frequency of the bar.

4. Magnetic Support and Superconducting Shield

The magnetic support system for the antenna consists of a set of persistent superconducting coils and a layer of superconducting Nb–Ti on the surface of the aluminum bar. Interaction of the current in the coils with image currents in the Nb–Ti provides the supporting force. Because the coils maintain constant magnetic flux linking them, increasing the height of the bar decreases the currents flowing in the support coils and consequently the supporting force until at the equilibrium height the support force balances the weight of the bar.

Extensive studies of the support system have been made on an aluminum bar 23 cm long and 15 cm in diameter. This bar and the support system are shown in Figure 3. Piezoelectric crystals are attached to the bar which is coated with a 0.04 cm layer of plasma-sprayed Nb–Ti. This assembly is inserted into a cradle which is fitted with a persistent levitation coil made of 0.06 cm × 0.13 cm copper-clad Nb–Ti wire. Another coil is mounted above the bar to simulate the effect of a heavier bar. These studies showed that the system was capable of supporting a 40 cm diameter bar with a considerable safety margin. Further, the mechanical Q of the bar while supported was observed to be unaffected by the Nb–Ti layer. At low temperature the observed Q in a vacuum environment was 6×10^5.

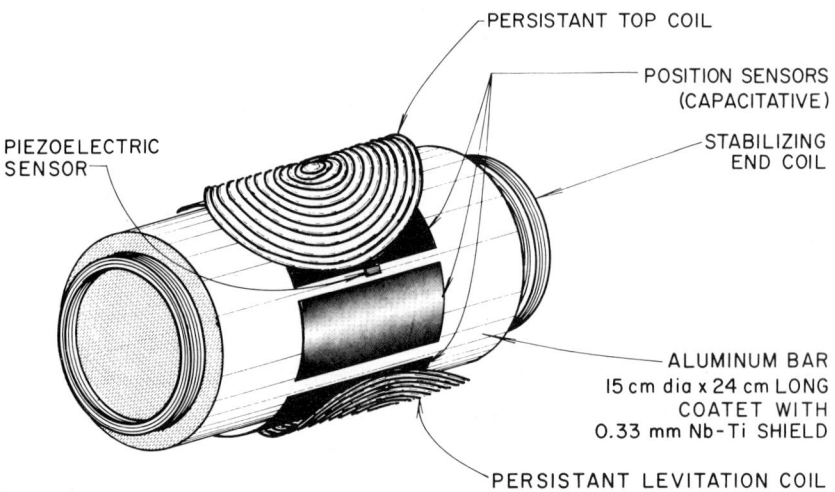

Fig. 3. Superconducting magnetic levitation system for test bar.

The field required to support a 40 cm diameter aluminum bar is about 2400 G; hence, the requirements on the superconducting coils are quite modest. We use 0.06 cm by 0.13 cm rectangular wire with a low field current capacity of 700 A, while only ~ 125 A are required to create a 2400 G field. For comparison a 1 m diameter bar requires a supporting field of 3400 G and a current of 170 A. Further experimentation with plasma spray technique has now enabled us to construct a shield which will exclude fields in excess of 3400 G and could thus be able to support a 1 m diameter bar.

The magnetic support has a natural resonant frequency associated with it which can be explained as follows. The magnetic flux is essentially trapped between the magnet coils and the surface of the bar. Any change in the height of the bar compresses the flux and increases the field strength at the surface of the bar. This in turn increases (or decreases) the force on the bar. The change in force F is $\Delta F \simeq F \Delta x/x$ where x is the distance from the bar to the coils. Hence there is an effective spring constant $K = F/x$. The resonant frequency is $\omega = (K/M)^{1/2} = (Mg/Mx)^{1/2} = (g/x)^{1/2}$. If $x \sim 1$ cm then $\omega \sim 30$ s^{-1} or $\omega/2\pi = 5$ Hz ($g =$ gravitational acceleration $\sim 10^3$ cm s^{-2}).

The magnetic support just described provides a number of important advantages over other means of support. Excellent mechanical isolation from vibrations at the 1350 Hz resonant frequency of the bar is achieved because of the low resonant frequency (approximately 5 Hz) of the support system and the absence of a direct acoustical path for higher frequency vibrations. Because the support forces are distributed along the bar, there is a relatively small coupling between the support coils and the bending modes of the bar. Distributing the support forces also minimizes acoustic emission arising from stresses induced in the bar by the support. Finally the magnetic support system provides thermal isolation to allow further cooling of the detector.

In addition to the magnetic support system, a cryogenic environment allows us to employ a superconducting Pb shield around the entire liquid helium dewar. This will provide excellent isolation from low frequency electromagnetic disturbances from outside the dewar. In particular this will make the antenna immune to changes in the Earth's ambient magnetic field. High frequency isolation is achieved automatically by the metal vacuum jacket of the entire system.

5. Cryogenic System

Figure 4 shows an end-on schematic view of a system to cool the prototype gravitational wave antenna to 1.2 K. Such a system is being assembled at Stanford.

The gravitational antenna consists of an aluminum bar 40 cm in diameter and 2 m long. The bar is coated with niobium titanium and floated on a superconducting magnet as was described in Section 4. The superconducting magnet is supported on a cylindrical aluminum container which contains liquid helium between two concentric aluminum cylinders. The helium container is supported by springs and acoustical filters in the vacuum system from an external beam which is supported by

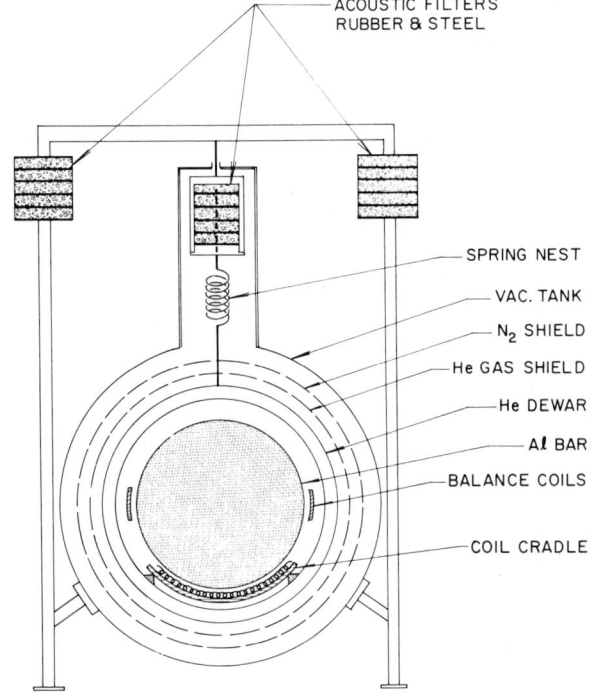

Fig. 4. Schematic of 1.2 K gravitational wave detector showing dewar and support system.

acoustic filters consisting of alternate layers of rubber and iron. The helium container is surrounded by a gas cooled shield which is cooled by the evaporating helium gas. This in turn is surrounded by a thermal shield cooled by liquid nitrogen. Between each set of shields is superinsulation consisting of several layers of aluminized mylar. Surrounding the helium dewar and in contact with it is a superconducting lead shield.

The dewar for this antenna system is 7 m long and 1 m in diameter on the outside and 60 cm in diameter on the inside. It is a modified superconducting accelerator dewar designed and built for the Stanford superconducting accelerator. Two gravitational wave antennas will be mounted in the single 7 m dewar.

Figure 5 is a schematic diagram of the ultra-low temperature part of the large 3 mK system which is being installed. The gravitational wave antenna consists of an aluminum bar 3 m long and 90 cm in diameter which weighs 5000 kg. This is coated with a superconducting niobium titanium layer about 0.4 mm thick. The bar is supported by a magnetic field of about 3400 G produced by a superconducting magnet. This provides an ideal isolation for vibration and for heat leaks. The bar is surrounded by a thermal shield cooled to 50 mK with a He^3–He^4 refrigerator. Surrounding the 50 mK shield is a shield cooled to 1.2 K by liquid helium. Surrounding the 1.2 K shield is a gas cooled shield, a liquid nitrogen cooled shield and a room temperature vacuum tank as shown in the 1.2 K detector in Figure 4.

The 5000 kg aluminum detector is cooled below 50 mK by a paramagnetic salt in the form of a toroid surrounded by a toroidal superconducting magnet and a superconducting shield. The magnet is turned on when the bar is in thermal contact with the He^4 bath. The heat of magnetization is removed at this temperature. He^4 exchange gas provides thermal contact between the He^4 bath and the He^3-He^4 cooled shield (50 mdeg shield). When it is desired to cool the 50 mK shield below the He^4 bath temperature the He^4 exchange gas is pumped out. He^3 exchange gas provides

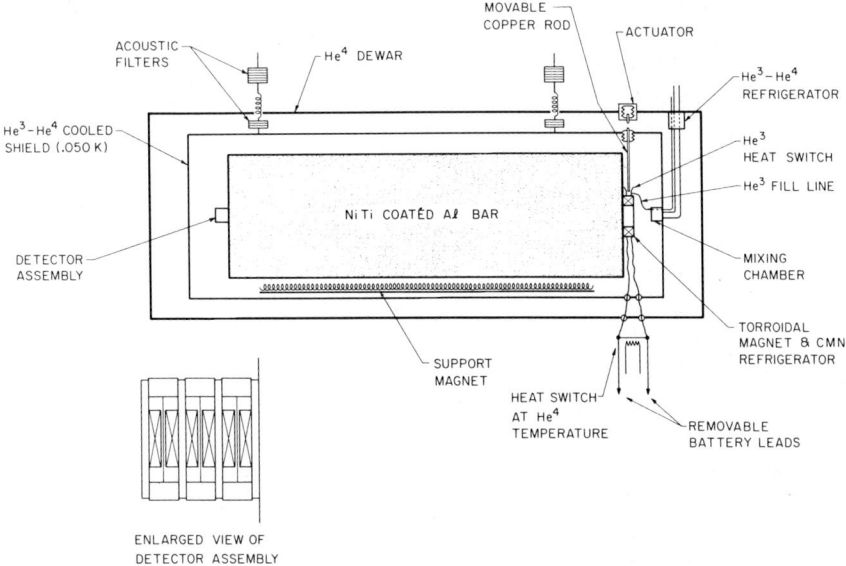

Fig. 5. Schematic of 3 mK gravitational wave detector and ultra low temperature part of dewar system.

thermal contact between the 50 mK shield and the aluminum bar. When the shield is cooled below 0.2 K by the He^3-He^4 refrigerator the He^3 exchange gas condenses out providing an effective heat switch. In order to cool the aluminum bar and salt below 0.2 K a mechanical He^3 heat switch is used as shown in Figure 5. A copper rod connects the 50 mK shield with the paramagnetic salt via a pool of He^3 into which the rod is pushed. When the bar and salt are cooled to 50 mK, the copper rod is withdrawn from the He^3 providing an effective heat switch. To cool further the magnet is slowly degaussed by means of heating a switch in the He^4 bath on the superconducting magnet. This thermal switch allows the superconducting magnet to decay slowly, cooling the salt and the aluminum bar. The salt may consist of cerium magnesium nitrate thermally mixed with liquid He^3 or it may consist of solid metallic paramagnetic salt which can be soldered to the container for good thermal contact. The salt when it is cold provides a large thermal reservoir to keep the bar cold for a calculated time of the order of 1 month.

Calculations on the amount of refrigeration required to cool the 5000 kg gravitational wave antenna may be of interest. Using only the latent heat of the liquid nitrogen it takes about 7000 l of liquid nitrogen to cool the dewar and bar to 63 K, the triple point of nitrogen. To cool from 63 K to 4 K using the heat of vaporization of liquid helium plus the specific heat of the helium vapor assuming perfect heat transfer requires 600 l of liquid helium. To cool from 4 to 1 K requires approximately 1 l of liquid helium. To cool from 1 K to 50 mK requires the removal of 137 joules with He^3–He^4 refrigerator plus the heat required to cool the paramagnetic salt. The He^3–He^4 refrigerator is designed to remove 10^5 ergs s^{-1} at 1 K, 5.5×10^4 ergs s^{-1} at 0.5 K and 500 ergs s^{-1} at 50 mdeg. Therefore it will take a few hours to cool from 1 K to 50 mK. To cool the aluminum bar from 50 mK to 3 mK requires a negligible amount of paramagnetic salt, less than one gram. The large amount of cerium magnesium nitrate which will be used (the order of 20 kg) will provide a larger thermal reservoir to keep the bar cold for an extended period of time.

The calculated heat leak into the 3 mK detector is dominated by cosmic rays. Assuming the dewar is surrounded by 15 cm thick lead shield, the heat leak into the bar from cosmic rays is about 0.8 ergs s^{-1}. The heat leak from radiation from the 50 mK shield is approximately 10^{-6} ergs s^{-1} assuming an emissivity of 0.05. The calculated heat leak from the leads connecting the magnetometer and the superconducting magnet with the external system via the 50 mK shield is also negligible. Preliminary experiments with a superconducting support lead us to believe that the heat leak from the magnetic support will also be negligible. If this should turn out to be the case then the total heat leak will be less than 1 erg s^{-1}. Under these conditions 30 kg of cerium magnesium nitrate would provide a thermal reservoir of sufficient heat capacity to keep the antenna below 5 mK for the order of 1 month. At this point the superconducting magnet would be turned on, the heat of magnetization removed and the salt again demagnetized to 3 mK. The turn around time for this operation will be limited by the capacity of the He^3–He^4 refrigerator and should be only a few hours. The thermal time constant for the aluminum bar at 1 K and below is calculated to be only a few milliseconds. Thus thermal equilibrium times are not limited by the large antenna but by the thermal boundary resistances between the paramagnetic salt and the antenna. Either cerous magnesium nitrate immersed in liquid He^3 as a heat transfer agent or a metallic paramagnetic refrigerant such as praseodymium copper 6 ($PrCu_6$) which can be soldered for thermal heat transfer could be used.

In summary, it appears theoretically possible to cool even a large antenna of 5 tons to a few millidegrees and keep it cold for long periods of time providing the ideal vibration isolation required for the gravitational wave experiments is realized.

6. Conclusion

We have considered the advantages of cooling a gravitational wave antenna to very low temperatures and have shown that it is theoretically possible to increase the sensitivity by 10^6 over room temperature detectors, while at the same time achieving ideal electrical and mechanical isolation. We have outlined the technique for cooling and have attempted to show that cooling to a few millidegrees and maintaining that temperature appears completely feasible.

It appears possible with this increased sensitivity to observe predicted events with sufficient signal-to-noise resolution to enable phase measurements to be made. Such measurements made simultaneously at each of the three locations at which the antennas are being built would enable an interferometric determination of the direction of the source. This might make possible a correlation between gravitational events and astronomical events in the electromagnetic spectrum, providing a basis for a new gravitational wave astronomy.

Acknowledgements

The research described in this paper was supported in part by the National Science Foundation and the Air Force Office of Scientific Research. The large dewars for Louisiana State University and for Stanford University to house the 3 mK antennas have been constructed by the Mississippi Test Facility of the National Aeronautics and Space Administration. The aluminum for the dewars and gravitational wave antennas is being supplied at cost by the Kaiser Aluminum Company.

References

Fairbank, W. M., et al.: 1972, in B. Bertotti (ed.), *Proc. Varenna Conf.*, in press.
Giacconi, R., et al.: 1971–1972, series of articles in *Astrophys. J. Letters*. For a summary see e.g. Gursky, H.: 1972, in Cécile M. DeWitt and B. S. DeWitt (eds.), *Black Holes*, Gordon and Breach, New York.
Gibbons, G. and Hawking, S.: 1971, *Phys. Rev.* **D4**, 2191.
Giffard, R. P., et al.: 1972, *J. Low Temp. Phys.* **6**, 533.
Hewish, A., et al.: 1968, *Nature* **217**, 709.
Klebesadal, R. W., et al.: 1973, *Astrophys. J. Letters* **182**, L85.
Paik, H. J.: 1972, in B. Bertotti (ed.), *Proc. Varenna Conf.*, in press.
Ruffini, R. and Wheeler, J. A.: 1971, in V. Hardy and H. Moore (eds.), *The Significance of Space Research for Fundamental Physics*, ESRO Publication SP-52, Paris, p. 45.
Ruffini, R.: 1972, in Cécile M. DeWitt and B. S. DeWitt (eds.), *Black Holes*, Gordon and Breach, New York.
Silver, A. H. and Zimmerman, J. E.: 1967, *Phys. Rev.* **157**, 317.
Tyson, J. A.: 1971, *Sixth International Conference on Gravitation and Relativity*, Copenhagen.
Weber, J.: 1960, *Phys. Rev.* **117**, 306.
Weber, J.: 1961, *General Relativity and Gravitational Radiation*, Interscience Publishers, Inc., New York.
Weber, J.: 1970, *Phys. Rev. Letters* **24**, 276.
Weber, J.: 1972, in B. Bertotti (ed.), *Proc. Varenna Conf.*, in press.
Zimmerman, J. E.: 1972, *Cryogenics* **12**, 19.

OPTIMIZATION OF GRAVITATIONAL BURST DETECTORS USING PIEZOELECTRIC TRANSDUCERS

D. MAEDER

Département de Physique Nucléaire et Corpusculaire, Université de Genève, CH-1211 Genève 4, Switzerland

Abstract. A general approach to overall system optimization is developed using the concepts of mechanical and electrical signal-to-noise ratios (MSNR and ESNR). These are proportional to $\sqrt{Q_{sys}/n}$ and to $\sqrt{\mu n}$, respectively, where Q_{sys} = mechanical Q of the complete detector, μ = (transducer mass/metal mass), and n = resolving time in units of one-half of the detector fundamental period. The overall SNR becomes a maximum for $n = n^*$ = optimum resolving time; this procedure yields MSNR* = = ESNR* = $\sqrt{2}$ SNR* $\propto \sqrt{Q_{sys} \cdot \mu}$ whereas $n^* \propto \sqrt{Q_{sys}/\mu}$.

Application to 'strong-coupling' type antennae (such as divided-cylinder systems) gives a high SNR* which depends very little on μ if $\mu > 0.01$. Q-factor and coupling efficiency relations were checked for 22 kg-prototypes using $\mu = 0.26\%$, 0.9%, and 1.8%. Two new detector configurations are suggested: the 'bridged-tube' allows strong-coupling for very long detectors; the 'folded-tube' operates at lower frequencies for a given length.

Reference

Maeder, D. G.: 1973, 'Matching Conditions in the Design of Gravitational Detector Systems', to be published in Compte-rendus du Colloque International C.N.R.S. No. 220, *Ondes et radiations gravitationnelles*, Paris.

ANALYSIS OF GRAVITATIONAL-WAVE DETECTION EXPERIMENTS*

DOUGLAS M. EARDLEY, DAVID L. LEE, and ALAN P. LIGHTMAN

California Institute of Technology, Calif., U.S.A.

R. V. WAGONER

Stanford University, U.S.A.

and

CLIFFORD M. WILL

University of Chicago, U.S.A.

Abstract. The structure of weak, plane, null gravitational waves is obtained for any metric theory of gravity. In general, six polarization states are present, which reduce to three (spin 0, ± 2) if the theory is to be quantizable. Schemes for obtaining the polarization amplitudes, as well as the direction and velocity of a wave, are presented.

* *Phys. Rev. Letters* **30**, 889 (1973).

ELECTROMAGNETIC DETECTORS OF GRAVITATIONAL WAVES*

V. B. BRAGINSKY, L. P. GRISHCHUK, A. G. DOROSHKIEVICH,
Ya. B. ZEL'DOVICH, I. D. NOVIKOV, and M. V. SAZHIN

Institute of Applied Mathematics, Moscow, Academy of Sciences, U.S.S.R.

Our group is investigating highfrequency gravitational waves (GW). The most promising approach to detection and laboratory generation of such GW seems to be through the transformation of GW into electromagnetic waves (EMW), and the reverse process: EMW→GW. The effects are small of course.

The generation, EMW→GW, depends on the gravitational effect of the density of electromagnetic energy, which is equal to $(E^2+B^2)/8\pi c^2$ and is of order 10^{-12} g cm^{-3} for $B=10^5$ G. The detection depends on h – the GW perturbations of the metric. To obtain $h=1$ one needs an energy flux $W=c^5/G=10^{59}$ erg s^{-1}.

On the other hand, there are factors which multiply the effect and inspire some hope. They are the resonance and coherence of waves. Although we give no final answer, the situation (ignoring technical difficulties) seems better than it did some years ago. Gertsenstein (*JETP*, 1961) made an important contribution to the theory of generation process. He considered an EMW propagating through a constant magnetic field B_0, so that the magnetic field of the wave B is parallel to B_0. A rigourous treatment was given by Boccaletti, de Sabbata, Fortini, Gualdi in *Nuovo Cimento*. In what follows we don't write tensor indices (see this paper for such details).

Due to the equality of the propagation velocities of GW and EMW, an EMW generates a GW with the same wave vector **K** and frequency ω. This is called coherence. The amplitude of the GW is proportional to the interaction length l; the coefficient q of energy transformaty on is proportional to l^2:

$$q = \frac{W(h)}{W(B_0)} = \frac{GB_0^2 l^2}{c^4}.$$

The equations for reciprocal transformations in a constant magnetic field have a very similar appearance. We use quantities h' and B' to describe GW and EMW, normalised in order to obtain equal coefficients in the energy flux:

$$W_{(GW)} \text{ erg cm}^{-2} \text{ s}^{-1} = (h')^2$$
$$W_{(EMW)} \text{ erg cm}^{-2} \text{ s}^{-1} = (B')^2$$

The equations are similar:

$$\Box h' = qB', \quad \Box B' = qh'$$

---

* Summary of a paper under the same title available as preprint No. 56–za 1973g from the authors.

where $\Box$ is the d'Alambert operator. One can introduce 'normal waves' f and g which are uncoupled:

$$h' + B' = f, \quad h' - B' = g.$$

Then

$$\Box f = qf, \quad \omega^2 = c^2k^2 - q$$
$$\Box g = qg, \quad \omega^2 = c^2k^2 + q.$$

Oscillations of the form 100% GW→100% EMW are predicted. But the domain of one complete oscillation is enourmous: it is the gravitational radius corresponding to the mass density of the constant magnetic field. The transformation coefficient q is proportional to the wave vector $|\mathbf{k}|$ therefore no superlight velocity occurs in the dispersion equation:

$$\omega = ck \pm \frac{q}{2ck}, \quad \frac{\partial \omega}{\partial k} = c = \text{const}.$$

The numerical calculations are shown for laboratory conditions, for pulsars and so on:

TABLE I

	B_0	l	q
Laboratoty	10^5	10^3	10^{-33}
Pulsar	10^{13}	10^6	10^{-11}
Cosmology $z=0$	$\leqslant 10^{-7}$	10^{28}	10^{-7}
Cosmology $z=10^3$	10^{-1}	10^{22}	10^{-4}

The effects are meagre. The last entries are a guess for a 'magnetic Universe': the greatest imaginable homogeneous field is used, whose energy density is equal to that of 2.7 K blackbody radiation and changes according to the same law: $\varepsilon \sim B^2 \sim (1+z)^4$ with z – the cosmological redshift.

The last entry is promising, but the heretofore neglected interaction of EMW with electrons and atoms destroys the coherence: using the amended equation $\Box B' = qh' + rB'$ we see that the aformentioned effect fails to occur.

Now we consider the case of a closed resonator for EMW. A resonator may be used as a source as well as a detector of GW. The possible types of EMW are classified as a set of eigen-solutions with definite frequencies. An EMW in a resonator with frequency ω produces a GW with frequency 2ω. The state and phase of EMW oscillations in a system of resonators may be ajusted in such a way that the whole system is working coherently.

To describe a resonator as a detector consider a change of the EM field in the resonator due to a GW with a frequency ω_g. At first, we neglect damping. EM field equations in a resonator without GW are:

$$B = a_n(t) f_n(x); \quad \frac{d^2 a_n}{dt^2} = -\omega_n^2 a_n; \quad a_n = a_{n0} e^{-i\omega_n t}.$$

Due to a GW they become

$$\frac{d^2 a_n}{dt^2} = -\omega_n^2 a_n + \omega_n^2 b_n e^{-i\omega_n t}, \quad b_n \sim a_n h, \quad \omega_n = \omega_m \pm \Omega.$$

Notice that the equations are written for time dependent amplitudes. We will not dwell here on the underlying Maxwell equations in a space curved by the gravitational wave. Thus these considerations give results which must be multiplied by numerical coefficients of order unity. Sometimes these coefficients are zero – but these exceptional cases (and the corresponding forms and positions of the resonators) should be avoided.

The GW introduces a mixture of different modes of oscillations. Resonance occurs if the GW frequency is equal to the difference of two EMW frequences.

Two particular cases should be mentioned: (1) a static initial field, $\omega_m = 0$, $a_n = a_{n0}$; and (2) parametric resonance with one type of oscillation, where $\Omega = 2\omega_n$. For the first case, with zero initial wave amplitude a_n, we introduce the energy transformation coefficient q as the ratio of EMW energy gain to the GW energy from through the resonator. In this case we have:

$$a_n \sim t, \; E_n \sim t^2, \quad q = \frac{E_n(t)}{I_g S t} = \frac{G B_0^2 t}{c^2 \omega_n}.$$

Comparing with the open case, we see that the gain due to the resonator is equal to ct/l (it is a pity that no reflection occurs and that no resonator can be used for GW – otherwise, our problems would be solved!) Here t is the duration of GW action. By cooling the resonator to a low temperature, one can avoid the spontaneous birth of EMW photons (resonator excitation). Still, the time t in the formulae for q is bound due to losses in the resonator.

In principle the energy gain is augmented if the initial amplitude of the two EMW are nonzero. In this case ΔE of one of them is proportional to the first power of the small quantity h the amplitude of the GW. But now the energy gain ΔE must be measured with respect to the background of already excited oscillations. No net gain is achieved.

Particularly in the case of parametric resonance, the effect is proportional to h. We have

$$\omega_m = \tfrac{1}{2}\Omega, \quad a_n \sim t, \quad \Delta E = Et \sqrt{\frac{16\pi G I_g}{c^3}}.$$

One can exploit the amplitude change

$$\frac{da_n}{dt} = \dot{h}_{2\omega} a_n \sin\theta$$

or phase change

$$\frac{d\varphi_n}{dt} = \dot{h}_{2\omega} \cos\theta$$

by the choice of phase difference θ between GW and EMW.

Here also the problem occurs.

The best parameters feasible give again of the order of 10^{-5} photons during a 1000 s cycle. When the generator and detector are separated one meter and $a_n = 0$ at $t = t_0$. Thus without some supplementary idea the detection scheme does not work! And nobody knows if the new idea will employ an EMW resonator.

Interesting in principle, although not the best for energy gain, is the situation when the resonator is in the form of a unidimensional waveguide. In this case, neglecting dispersion, one can consider a wave packet with definite front and rear ends – 1 and 2. Geometrical optics can be used; one knows that the mode number in the region between 1 and 2 is constant; therefore, if a systematic change of the length $l_2 - l_1$ occurs, the frequency is shifted according to

$$\Delta\omega/\omega = \Delta l_{21}/l_{21}.$$

The case of a annular waveguide is typical. The propagation is along the φ coordinate; $r = $ const. The part of the metric with dt and $d\varphi$ (but $dr = 0$, $dz = 0$) is

$$ds^2 = c^2\,dt^2 - r^2(1 - h_{22})\,d\varphi^2.$$

The metric perturbation due to a circularly polarized GW is included. The resonance case occurs if motion of packet is always in phase with the metric distortion. In this case the packet length or frequency depends on time linearly.

Another treatment of the problem could be given by decomposing a wave of finite length a superposition of elementary eigenoscillations, with

$$\omega_{n+1} = \omega_n + \frac{c}{r}.$$

The frequency shift due to GW is smaller than the frequency difference of two adjacent eigenoscillations. Therefore the action of GW could be described as the transfer of energy from one mode to another.

The two treatments are equivalent. What is worth mentioning in the geometrical optics approach is the selection rules. A straight waveguide with mirrors on the ends gives no systematical effects if it is orientes along the propagation direction of the GW. This is due to the transverse character of GW. But the straight waveguide also does not work in the perpendicular plane and this is a non trivial selection rule. One must go to an annular waveguide or incline the straight waveguide. But here we are going into details important for obtaining the best gain from resonators. This has meaning only in the case when order of magnitude estimates suggest that on experiment is possible, which is unfortunately not yet the case.

So it is appropriate to end the discussion with the slogan 'New ideas are badly needed'.

INTERACTION OF GRAVITATIONAL RADIATION WITH A UNIFORMLY MAGNETIZED SPHERE

V. DE SABBATA, P. FORTINI, and C. GUALDI

Istituto di Fisica dell'Università Bologna, Bologna, Italy

and

L. FORTINI BARONI

Holder of a C.N.R. 'Italian National Research Council' Scholarship, Bologna, Italy

Abstract. Maxwell equations in the field of a gravitational wave are linearized by means of the weak field approximation. Then the equations are solved in the case of a uniformly magnetized sphere and the dipole electromagnetic radiation power is calculated. These results are applied to compute the electromagnetic radiation emitted by magnetic neutron stars and by the Earth when hit by gravitational radiation.

GRAVITATIONAL RADIATION BY ULTRARELATIVISTIC BODIES

PETER JOCELYN WESTERVELT

University of Texas, Austin, Tex. 78712, U.S.A.

and

Brown University, Providence, R.I. 02912, U.S.A.

Abstract. I have shown (Westervelt, 1966) that ultrarelativistic bodies do not radiate gravitational waves in the forward direction. This work has been extended so as to apply to circular orbits. Even if low efficiency of generation precludes direct observation of gravitational waves, indirect evidence for their existence is available in a recent analysis (Westervelt, 1969) of Shapiro's fourth test of general relativity.

References

Westervelt, P. J.: 1966, *J.E.T.P. Letters* **4**, 225.
Westervelt, P. J.: 1969, *Acta Phys. Polon.* **35**, 203.

PART II

STABILITY AND COLLAPSE

THE STABILITY OF RELATIVISTIC SYSTEMS

S. CHANDRASEKHAR

University of Chicago, Chicago, Ill., U.S.A.

Abstract. The stability of relativistic systems is reviewed against the background of what is known in the corresponding contexts of the Newtonian theory. In particular, the importance of determining whether Dedekind-like points of bifurcation occur along given stationary axisymmetric sequences is emphasized: the occurrence of such points of bifurcation may signal the onset of secular instability induced by radiation-reaction. (At a Dedekind-like point of bifurcation, the system can be subject, quasistationarily, to a non-axisymmetric deformation with an $e^{2i\phi}$-dependence on the azimuthal angle ϕ.)

A formalism is described in terms of which the normal modes of axisymmetric oscillation of axisymmetric systems can be determined. Specialized to neutral modes of oscillation the formalism provides an alternative proof of Carter's theorem and clarifies the minimal requirements for its validity. A parallel formalism is described for ascertaining whether an axisymmetric system can be subject to a quasi-stationary non-axisymmetric deformation. The possibility of applying this latter formalism to determining whether a Dedekind-like point of bifurcation occurs along the Kerr sequence is considered.

1. Introduction

During recent years the stability of a variety of systems in general relativity has been considered by a number of investigators by differing methods. In this paper, an approach to a substantial class of them will be presented which will make them appear as special cases of an effectively single mathematical theory. Moreover, the approach will be motivated by an attempt to develop the problems in general relativity closely parallel to the corresponding developments in the Newtonian theory.

The systems whose stability we shall consider fall into two categories: those which are spherically symmetric and those which are axisymmetric (by virtue of rotation, for example). In the case of spherical systems, we shall be interested in both radial and non-radial oscillations and in the instabilities which derive from them; and in the case of axisymmetric systems we shall be interested in oscillations which preserve the axisymmetry and in those which do not. And in all cases, we shall be interested in criteria for the occurrence of neutral modes of oscillation since their occurrence often signals the onset of instabilities.

2. Known Results in the Newtonian Theory

We shall first enumerate the known results of stability analysis in the Newtonian theory.

With respect to spherical systems, it is known that stability with respect to radial oscillations depends on an average value of the adiabatic exponent, γ, which relates the Lagrangian changes, Δp in the pressure and $\Delta \varrho$ in the density, which a fluid element experiences during its motion:

$$\Delta p/p = \gamma \, \Delta \varrho/\varrho. \tag{1}$$

Thus, instability will occur if

$$\bar{\gamma} = \frac{\int_0^M \gamma p \, \mathrm{d}M(r)}{\int_0^M p \, \mathrm{d}M(r)} < \tfrac{4}{3} \quad [\mathrm{d}M(r) = 4\pi r^2 \varrho \, \mathrm{d}r]; \tag{2}$$

and the *e*-folding time of the instability (when it occurs) is of the same order as the period of pulsation (when the configuration is stable).

The *dynamical instability* to radial perturbations which occurs when $\bar{\gamma} < \tfrac{4}{3}$ is a *global instability* in the sense that its occurrence depends on the structure of the configuration in its entirety. It is a remarkable fact that there are no other global instabilities to which a spherical distribution of mass is subject: instabilities that may arise from non-radial perturbations are all *local* and originate (as Lebovitz first rigorously showed) in the violations of the Schwarzschild criterion for convective stability.

When we consider rotating systems which are axially symmetric in the stationary state, the instabilities that may arise are of two kinds. First, the continuation into the rotating domain of the global instability that occurs in non-rotating systems for radial perturbations; and second, instabilities that derive from the centrifugal and coriolis forces that are operative in rotating systems, i.e. in instabilities that are peculiar to rotating systems.

It is known (cf. Lebovitz, 1970) that the former type of instabilities result from modes of axisymmetric oscillations; indeed, for slow rotation, the condition (2) is replaced by

$$\bar{\gamma} - \tfrac{4}{3} + \text{constant} \frac{\Omega^2}{\pi G \bar{\varrho}} < \tfrac{4}{3}, \tag{3}$$

where $\bar{\varrho}$ is the mean density of the configuration. More generally, an exact expression can be given for the fundamental frequency (σ) of axisymmetric oscillation of a slowly rotating configuration in terms of the frequency (σ_0) of radial oscillation of the non-rotating configuration and the ($l=0$)-distortion of it by the rotation. Thus,

$$\sigma^2 = \sigma_0^2 + \Omega^2 \sigma_1^2 + O(\Omega^4), \tag{4}$$

where σ_1^2 depends only on the proper solution ξ_0 belonging to σ_0 and the spherically symmetric part of the distortion caused by the rotation.

With regard to the latter type of instabilities, i.e. those that are peculiar to rotating systems, they derive from non-axisymmetric perturbations; and these instabilities are of two kinds: *secular* and *dynamical*. We shall clarify their nature and their origin by considering the classical sequences of uniformly rotating homogeneous masses. (For more detailed information on what follows, see Chandrasekhar, 1969).

It is well known that a possible sequence of equilibrium figures for rotating homogeneous masses is represented by the Maclaurin sequence of oblate spheroids. When one examines the second-harmonic oscillations of the Maclaurin spheroid, in a frame of reference rotating with its angular velocity, one finds that for two of these modes, whose dependence on the azimuthal angle is given by $e^{2i\phi}$, the squares of the characteristic frequencies depend on the eccentricity e in the manner illustrated in Figure 1. It will be observed that one of the modes becomes neutral (i.e. $\sigma^2 = 0$) when $e = 0.813$ and that the two modes coincide when $e = 0.953$ and become complex conjugates of one another beyond this point. Accordingly, the Maclaurin spheroid becomes *dynamically unstable* at the latter point (first isolated by Riemann). On the other hand, the origin of the neutral mode at $e = 0.813$ is that at this point a new equilibrium sequence of tri-axial ellipsoids – the ellipsoids of Jacobi – bifurcate. On this latter account it was conjectured by Lord Kelvin in 1883 that "if there be any viscosity, however slight, in the liquid, or if there be any imperfectly elastic solid, however small, floating on it or sunk within it, the equilibrium [beyond $e = 0.81$] cannot be secularly stable." Lord Kelvin's conjecture has been confirmed by an explicit investigation of the effect of a small amount of viscous dissipation on the two modes illustrated in Figure 1. It is found that viscous dissipation makes the mode, which becomes neutral at $e = 0.81$, unstable beyond this point with an e-folding time which depends inversely on the magnitude of the kinematic viscosity and which further decreases monotonically to zero at the point ($e = 0.953$) of onset of dynamical instability. (The e-folding time of the instability becomes proportional to $v^{1/2}$ in the immediate neighborhood of $e = 0.95$.)

However, it should not be concluded that *any* dissipative mechanism will make the Jacobi mode unstable beyond the point of bifurcation. If we ask, for example, what effect the dissipative forces derived from radiation-reaction of general-relativistic origin, has on the secular stability of the Maclaurin spheroid at $e = 0.813$, we find (Chandrasekhar, 1970) that it does *not* induce any stability in the Jacobi mode; instead, it induces instability in the *alternative* mode at the same eccentricity. In the first instance this may seem surprising; but the situation we encounter here clarifies some important issues.

If instead of analyzing the normal modes in the rotating frame, we had analyzed them in the inertial frame, we should have found that the mode, which becomes unstable by radiation-reaction at $e = 0.813$, is in fact neutral at this point. And the neutrality of *this* mode in the inertial frame corresponds to the fact that the neutral deformation is associated with the bifurcation at this point of a new tri-axial sequence – the sequence of Dedekind ellipsoids. These Dedekind ellipsoids, while they are congruent to the Jacobi ellipsoids, differ from them in that they are at rest in the inertial frame and owe their tri-axial figures to internal vortical motions. An important conclusion that would appear to follow from these facts is that in the framework of general relativity we can expect secular instability, derived from radiation-reaction, to arise by a Dedekind mode of deformation (which is quasi-stationary in the inertial frame) rather than by a Jacobi mode (which is quasi-stationary in a rotating frame). A further fact which requires to be emphasized in this context is that the notion of a neutral point is subject to ambiguity arising from the freedom we have in the choice of a coordinate frame in which we may wish to specify the characteristic frequencies belonging to the various normal modes. It is important to observe in this connection that while for uniformly rotating objects, the inertial frame and the frame rotating with the object are naturally distinguished, no rotating frame is naturally distinguished when the object is rotating non-uniformly. Accordingly, in the case of non-uniform

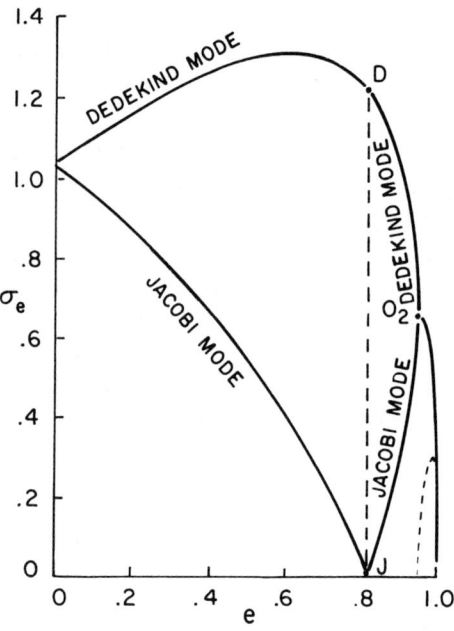

Fig. 1. The characteristic frequencies (in the unit $(\pi G\varrho)^{1/2}$) of the two even modes of second-harmonic oscillation of the Maclaurin spheroid. The Jacobi sequence bifurcates from the Maclaurin sequence by the mode that is neutral ($\sigma^2 = 0$) at $e = 0.813$; and the Dedekind sequence bifurcates by the alternative mode at D. At O_2 ($e = 0.9529$) the Maclaurin spheroid becomes dynamically unstable. The real and the imaginary parts of the frequency, beyond O_2, are shown by the full-line and the dashed curves, respectively. Viscous dissipation induces instability in the branch JO_2 of the Jacobi mode; and radiation-reaction induces instability in the branch DO_2 of the Dedekind mode.

rotation, the one secure concept is that of dynamical instability. The points at issue here are important particularly in view of differentially rotating compressible models that have recently been constructed by Ostriker and his associates (Tassoul and Ostriker, 1968; Ostriker and Tassoul, 1969; Ostriker and Bodenheimer, 1973; Durisen, 1973a, b) and their demonstration that these sequences have extraordinary similarity with the classical Maclaurin sequence. These investigations on differentially rotating systems would lead one to suppose that during the natural evolution of rotating systems, disc-like objects cannot come into being by virtue of secular or dynamical instability setting in long before the objects become anything like disc-like (cf. Chandrasekhar and Lebovitz, 1973; Chandrasekhar, 1974).

3. Known Results in General Relativity

The stability of spherically symmetric fluid masses in general relativity is one of the first problems that was fully and rigorously solved (Chandrasekhar, 1964; Fowler, 1964). And the theory parallels the Newtonian theory very closely. Thus, the criterion for instability is again an inequality for an average value of γ though it cannot in general be expressed as simply as in Equation (2). By actual numerical integrations (Bardeen et al., 1966; Thorne and Meltzer, 1966) of the pulsation equation it has been shown that instability can occur under circumstances in which the Newtonian theory will predict stability. This *relativistic instability* is of particular importance when γ is close to, but greater than $\frac{4}{3}$, as it will be the case when radiation pressure is dominant (as in massive stars) or when the constituent particles that contribute to the pressure move with velocities close to the velocity of light (as in degenerate configurations near their limiting mass). Thus, if γ should be a constant through the star and exceeds $\frac{4}{3}$ by a small amount, then it follows from the theory that instability will occur when the radius of the star

$$R < \frac{2GM}{c^2} \frac{K}{\gamma - \frac{4}{3}}, \tag{5}$$

where K is a constant which, while it depends on the structure of the star, is generally of order unity. This last formula shows very clearly that effects arising from general relativity can induce instability even under circumstances when its effect on the structure of the equilibrium configuration are entirely negligible.

The corresponding theory of axisymmetric oscillations of slowly rotating stars has been recently worked out by Hartle and Thorne (1968) and by Chandrasekhar and Friedman (1972b). Again this theory parallels the Newtonian theory; and a formula identical in form with Equation (4) exists in the theory of general relativity as well.

The principal reason why the theory of radial oscillations of spherical systems and of axisymmetric oscillations in slowly rotating systems, in general relativity, closely parallel the Newtonian theory is that in both cases gravitational radiation plays no role: it is identically absent in the former case (by virtue of Birkhoff's theorem) and it is absent to the relevant order in the latter case. But this simplification does not obtain for non-radial oscillations of spherical systems and axisymmetric oscillations of rapidly rotating systems. We now turn our consideration to such systems.

4. A General Theory of Axisymmetric Oscillations

It is clear that the theory of radial oscillations of spherical systems can be included as a special case of a general theory of axisymmetric oscillations of axisymmetric systems. The theory of non-radial oscillations of spherical systems can also be included as a special case of the same general theory since the normal modes of such oscillations can be analyzed in spherical harmonics $Y_l^m(\theta, \phi)$; and while the characteristic frequencies belonging to the normal modes will depend on l they will be independent of m. In other words, the radial and the non-radial oscillations of spherical systems as well as the axisymmetric oscillations of rotating axisymmetric systems can all be included in one general theory. We shall indicate later (in Section 6 below) how the same theory with changes only in notation can be adapted to isolate points of onset of instabilities by non-axisymmetric modes of oscillation.

The theory of non-radial oscillations of spherical systems with the emission of gravitational radiation has been developed in considerable detail by Thorne and his associates (Thorne and Campolattaro, 1967, 1970; Thorne, 1969a, b; Thorne and Price, 1969) and more recently by Ipser and Detweiller (1973). The theory we shall outline includes the theory of non-radial oscillations in an alternative version.

The account which follows is largely based on a series of papers by Chandrasekhar and Friedman that have been published during the past two years (Chandrasekhar and Friedman 1972a, b, 1973a, b, c).

We start then with a form for the metric that is suitable to describe general time-dependent systems with the only restriction that the systems maintain axial symmetry, about a fixed axis, at all times. It can be shown that a form of the metric that is suitable for the purposes is

$$ds^2 = -e^{2\nu}(dt)^2 + e^{2\psi}(d\phi - q_{2,0}\,dx^2 - q_{3,0}\,dx^3 - \omega\,dt)^2 + \\ + e^{2\mu_2}(dx^2)^2 + e^{2\mu_3}(dx^3)^2, \tag{6}$$

where $\nu, \psi, \omega. q_2, q_3, \mu_2$, and μ_3 are functions of the time coordinate $t(=x^0)$ and the two space-like variables x^2 and x^3 but independent of the cyclic angle-variable $\phi(=x^1)$. As we have written, the metric depends on the seven functions enumerated; but it can be shown that the functions q_2, q_3, and ω can occur in the field equations only in the combinations

$$\omega_{,2} - q_{2,00}, \quad \omega_{,3} - q_{3,00}, \quad \text{and} \quad q_{2,30} - q_{3,20}; \tag{7}$$

accordingly, we shall be concerned with only six independent quantities.

The metric (6) includes the form.

$$ds^2 = -e^{2\nu}(dt)^2 + e^{2\psi}(d\phi - \omega\,dt)^2 + e^{2\mu_2}(dx^2)^2 + e^{2\mu_3}(dx^3)^2, \tag{8}$$

which is generally chosen as appropriate for stationary axisymmetric systems. However, in the stationary case, when ν, ψ, ω, μ_2, and μ_3 are functions of x^2 and x^3 only, one has the additional gauge freedom to restrict the functions μ_2 and μ_3 by a coordinate condition, such as $\mu_2 = \mu_3 = \mu$ or $e^{\mu_3} = x^2 e^{\mu_2}$. In the non-stationary case, we do not have this freedom. Thus, if we should consider time-dependent departures from equilibrium, we must allow for a difference in the Eulerian changes in μ_2 and μ_3.

In writing out the field equations appropriate to the metric (6), we shall restrict ourselves to the case when the source of the gravitational field is a perfect fluid described by the energy-momentum tensor

$$T^{ij} = (\varepsilon + p)\,u^i u^j + p g^{ij}, \tag{9}$$

where ε and p denote the energy density and the pressure, respectively. We shall further suppose that there exists an 'equation of state' which relates ε uniquely as a function of p and the baryon number N (per unit proper three-volume):

$$\varepsilon = \varepsilon(p, N). \tag{10}$$

The equations of the problem are then provided by the Einstein field-equations supplemented by the equation

$$(Nu^j \sqrt{-g})_{,j} = 0 \tag{11}$$

which ensures the conservation of the baryon number.

Since we have assumed that the system preserves its axisymmetry at all times, we shall have in addition to Equation (11), the further equation

$$u^j \left(\frac{\varepsilon + p}{N} u_1 \right)_{,j} = 0 \tag{12}$$

which ensures the conservation of the angular momentum per baryon.

The problem to which we now address ourselves is the following. We are given a system that is axisymmetric, stationary and in a state of uniform rotation with an angular velocity Ω. We suppose that it is subjected to an infinitesimal perturbation and that in the non-stationary state which ensues, the axisymmetry of the system is maintained. What are the equations which govern the time evolution of such a perturbation?

We start with the full, non-linear, time-dependent equations that are appropriate to metric (6). The initial stationary system will satisfy these same equations: only the terms which explicitly involve the time derivatives will be absent. The equations which govern the perturbation can be obtained by linearizing the full time-dependent equations about the time-independent solution representing the initial stationary state.

In considering the changes in the various quantities caused by the perturbation, we shall distinguish between the Eulerian and the Lagrangian changes. These are respectively the changes that take place at a fixed location and the changes that accompany a fluid element as it moves. It is convenient to describe the perturbations in the various quantities that describe the fluid in terms of a Lagrangian displacement ξ which is the spatial displacement that an element of fluid experiences relative to its location in the unperturbed state. Since we have assumed that the perturbation does not affect the axisymmetry of the configuration, it is clear that the components ξ^α ($\alpha = 2, 3$) of ξ should suffice to describe the displacement. The Lagrangian displacement ξ^α is related to the velocity v^α ($= dx^\alpha/dt$) by

$$v^\alpha = \xi^\alpha_{,0}. \tag{13}$$

Among the equations which govern the departures from equilibrium of a system that is initially static or stationary we distinguish two classes: *initial-value equations* and *dynamical equations*. Initial-value equations are those that are of the first order in the time derivatives; and dynamical equations are those that are of second order in the time derivatives. Initial value equations can be directly integrated when the fluid variables are expressed in terms of the Lagrangian displacement. In contrast, dynamical equations are the basic equations of motion that eventually lead to a characteristic-value problem which determines the normal modes of oscillation and the characteristic frequencies belonging to them.

For the problem we are presently considering, the initial value equations are (1) the equations expressing the conservation of baryon number

$$\frac{\Delta N}{N} = -\frac{1}{u^0\sqrt{-g}}(\xi^\alpha u^0\sqrt{-g})_{,\alpha} - \delta[\log(u^0\sqrt{-g})] =$$

$$= -\frac{1}{u^0\sqrt{-g}}(\xi^\alpha u^0\sqrt{-g})_{,\alpha} - \frac{V\delta V}{1-V^2} - \delta(\psi + \mu_2 + \mu_3); \tag{14}$$

(2) the adiabatic condition expressing the conservation of entropy

$$\frac{\Delta p}{p} = \gamma \frac{\Delta N}{N} \quad \text{and} \quad \Delta\left(\frac{\varepsilon + p}{N}\right) = vp\frac{\Delta N}{N^2}; \tag{15}$$

(3) the equation expressing the conservation of angular momentum

$$\frac{\delta V}{V(1-V^2)} = -\frac{\gamma p}{\varepsilon + p}\frac{\Delta N}{N} - \delta\psi - \xi^\alpha(\log u_1)_{,\alpha}; \tag{16}$$

and finally (4) the $(0, 2)$- and the $(0, 3)$-components of the linearized field equations

$$(\delta\psi + \delta\mu)_{,2} - v_{,2}(\delta\psi + \delta\mu) + \psi_{,2}(\delta\psi - \delta\mu) =$$
$$= e^{2\mu_2}\left(8\pi\frac{\varepsilon + p}{1-V^2}\xi^2 - \frac{Q}{2\sqrt{-g}}\omega_{,3}\right) - \delta\tau_{,2} - (2\mu_3 + \psi - v)_{,2}\,\delta\tau \tag{17}$$

and

$$(\delta\psi + \delta\mu)_{,3} - v_{,3}(\delta\psi + \delta\mu) + \psi_{,3}(\delta\psi - \delta\mu) =$$
$$= e^{2\mu_3}\left(8\pi\frac{\varepsilon + p}{1-V^2}\xi^3 + \frac{Q}{2\sqrt{-g}}\omega_{,2}\right) + \delta\tau_{,3} + (2\mu_2 + \psi - v)_{,3}\,\delta\tau, \tag{18}$$

where

$$u^0 = \frac{e^{-v}}{\sqrt{(1-V^2)}}, \quad u_1 = \frac{e^\psi V}{\sqrt{(1-V^2)}}, \quad V = e^{\psi - v}(\Omega - \omega),$$

$$\delta\mu = \tfrac{1}{2}\delta(\mu_3 + \mu_2), \quad \delta\tau = \tfrac{1}{2}\delta(\mu_3 - \mu_2), \tag{19}$$

and

$$Q = e^{3\psi + v - \mu_2 - \mu_3}(q_{2,3} - q_{3,2}). \tag{20}$$

The $(1, 2)$- and the $(1, 3)$-components of the linearized field-equations play a double role in the theory: they provide initial-value equations of $\delta\omega$ while their integrability condition leads to a dynamical equation for Q. Thus

$$\delta\omega_{,2} - q_{2,00} = 16\pi(\varepsilon+p)\, u^0 u_1 e^{-2\psi+2\nu+2\mu_2}\xi^2 - \\ - \omega_{,2}(3\delta\psi - \delta\nu + 2\delta\tau) - e^{-3\psi+\nu+\mu_2+\mu_3} Q_{,3} \quad (21)$$

and

$$\delta\omega_{,3} - q_{3,00} = 16\pi(\varepsilon+p)\, u^0 u_1 e^{-2\psi+2\nu+2\mu_3}\xi^3 - \\ - \omega_{,3}(3\delta\psi - \delta\nu - 2\delta\tau) + e^{-3\psi+\nu+\mu_2+\mu_3} Q_{,2} ; \quad (22)$$

and the elimination of $\delta\omega$ from these equations gives

$$(e^{-3\psi+\nu+\mu_3-\mu_2} Q_{,2})_{,2} + (e^{-3\psi+\nu+\mu_2-\mu_3} Q_{,3})_{,3} = e^{-3\psi-\nu+\mu_2+\mu_3} Q_{,00} - \\ - [\omega_{,2}(3\delta\psi - \delta\nu + 2\delta\tau)]_{,3} + [\omega_{,3}(3\delta\psi - \delta\nu - 2\delta\tau)]_{,2} + \\ + 16\pi\{[(\varepsilon+p)\, u^0 u_1 e^{-2\psi+2\nu+2\mu_2}\xi^2]_{,3} - \\ - [(\varepsilon+p)\, u^0 u_1 e^{-2\psi+2\nu+2\mu_3}\xi^3]_{,2}\}, \quad (23)$$

which, as we have stated, is a dynamical equation for Q.

Turning to the remaining dynamical equations, we shall separate the time and the space variables and seek solutions whose dependence on time is given by

$$e^{i\sigma t}, \quad (24)$$

where σ is a characteristic-value parameter to be determined. This time-dependence will appear as a factor in all the equations. We shall suppose that this common factor has been removed and that all quantities (such as ξ^α, ΔN, etc.) which appear in the equations from now on represent the space-dependent amplitudes of the respective quantities; thus it will be assumed, for example, that the chosen Lagrangian displacement is of the form

$$\xi^\alpha(x^2, x^3)\, e^{i\sigma t}. \quad (25)$$

The dynamical equation which follows from the linearization of the equation

$$u^j u_{i;j} = -\frac{1}{\varepsilon+p}(\delta^j{}_i + u^j u_i)\, p_{,j}, \quad (26)$$

(which in turn is a consequence of the identity $T^{ij}{}_{;j}=0$) is the pulsation equation:

$$-\sigma^2(\varepsilon+p)(u^0)^2\, e^{2\mu_\alpha}\xi^\alpha = -u^0 \left(\frac{\gamma p}{u^0}\frac{\Delta N}{N}\right)_{,\alpha} + \frac{\Delta N}{N} p_{,\alpha} + \\ + (\varepsilon+p)\left(\frac{\Delta u^0}{u^0}\right)_{,\alpha} - \\ - (\varepsilon+p)\, u^0 u_1 (\Delta\Omega_{,\alpha} - q_{\alpha,00}). \quad (27)$$

Besides this pulsation equation we must include, at most, two of the linearized field equations; 'at most two' since Equations (17), (18), 21), and (22) already account for four equations and there can be no more than six linearly independent field equations.

As the remaining dynamical equations we may take, for example, the (1, 1)- and the [(2, 2)+(3, 3)]-components of the field equations. We shall not write these equations out explicitly.

Our problem is to solve Equation (27) consistently with the initial-value Equations (14)–(18), (21), and (22), the remaining dynamical equations, and the appropriate boundary conditions. The boundary conditions are that ξ^α vanishes at the origin and remains bounded and continuous over its domain; that Δp vanishes on the boundary and that all the remaining field variables (such as $\delta\psi$, δv, etc.,) vanish sufficiently rapidly at infinity and satisfy the necessary conditions on the horizon if we are dealing with vacuum solutions exterior to a black hole.

The problem to which we are thus led is a characteristic value problem for σ^2.

5. The Variational Principle: Applications to Vacuum Metrics

The characteristic-value problem to which we were led in Section 4 can be formulated in terms of a variational principle. We shall clarify its nature by restricting ourselves to the case of vacuum metrics external to an event horizon and which are asymptotically flat at infinity, i.e. effectively to the Kerr and the Schwarzschild metrics.

First by making use of the equations governing the problem (it may be noted parenthetically that the initial-value equations and the dynamical equations must be treated differently) we derive the following formulae for σ^2.

$$\sigma^2 \int\int [\![2e^{-2v}\sqrt{-g}\{(\delta\tau)^2+(\delta\psi)^2-[\delta(\psi+\mu)]^2\}+$$
$$+\tfrac{1}{2}e^{-3\psi-v+\mu_2+\mu_3}Q^2]\!] \, dx^2 \, dx^3 =$$
$$= \int\int [\![X \, e^{3\psi-v}[4(\delta\psi)^2+(\delta\tau)^2]+4Y \, e^{3\psi-v} \, \delta\psi\delta\tau -$$
$$-4U(\delta\tau)^2+2e^\beta [e^{\mu_3-\mu_2}(\delta\psi_{,2})^2+e^{\mu_2-\mu_3}(\delta\psi_{,3})^2]-$$
$$-4e^\beta \{e^{\mu_3-\mu_2}[\beta_{,2}\delta\mu_{3,2}-(\psi-v)_{,2} \, \delta\psi_{,2}]-$$
$$-e^{\mu_2-\mu_3}[\beta_{,3}\delta\mu_{2,3}-(\psi-v)_{,3} \, \delta\psi_{,3}]\}\,\delta\tau -$$
$$-2[2(Q_{,2}\omega_{,3}-Q_{,3}\omega_{,2})\,\delta\psi -(Q_{,2}\omega_{,3}+Q_{,3}\omega_{,2})\,\delta\tau]+$$
$$+\tfrac{1}{2}e^{-3\psi+v}[e^{\mu_3-\mu_2}(Q_{,2})^2+e^{\mu_2-\mu_3}(Q_{,3})^2]]\!] \, dx^2 \, dx^3 +$$
$$+\int [\![e^{\mu_3-\mu_2+\beta}[-\delta(\psi-v)_{,2}\,\delta\psi -2(\psi-v)_{,2}\,\delta\psi\delta\tau +$$
$$+\delta(\psi+v)_{,2}\,\delta\mu_3+2(\psi+v)_{,2}\,\delta\tau\delta\mu_3]+$$
$$+Q\omega_{,3}\delta(2\psi-\tau)-\tfrac{1}{2}e^{-3\psi+v+\mu_3-\mu_2}QQ_{,2}]\!]_{[x^2]} \, dx^3 +$$
$$+\int [\![e^{\mu_2-\mu_3+\beta}[-\delta(\psi-v)_{,3}\,\delta\psi +2(\psi-v)_{,3}\,\delta\psi\delta\tau +$$
$$+\delta(\psi+v)_{,3}\,\delta\mu_2-2(\psi+v)_{,3}\,\delta\tau\delta\mu_2]-$$
$$-Q\omega_{,2}\delta(2\psi+\tau)-\tfrac{1}{2}e^{-3\psi+v+\mu_2-\mu_3}QQ_{,3}]\!]_{[x^3]} \, dx^2. \qquad (28)$$

where

$$\beta = \psi + v,$$
$$X = e^{\mu_3 - \mu_2}(\omega_{,2})^2 + e^{\mu_2 - \mu_3}(\omega_{,3})^2,$$
$$Y = e^{\mu_3 - \mu_2}(\omega_{,2})^2 - e^{\mu_2 - \mu_3}(\omega_{,3})^2$$

and

$$U = e^{\beta}[e^{\mu_3 - \mu_2}(\beta_{,2}\mu_{3,2} + \psi_{,2}v_{,2}) + e^{\mu_2 - \mu_3}(\beta_{,3}\mu_{2,3} + \psi_{,3}v_{,3})]. \tag{29}$$

Also, in equation (28) the symbol

$$[\![\ldots]\!]_{[x^\alpha]}$$

in the integrands of the surface integrals has the following meaning. For a fixed x^{β} ($\beta \neq \alpha$) let the appropriate limits of x^{α} be $x^{\alpha}(1)$ and $x^{\alpha}(2) \geqslant x^{\alpha}(1)$; the symbol stands for the difference in the values of the quantity enclosed by the double brackets at $x^{\alpha}(2)$ and $x^{\alpha}(1)$.

Equation (28) provides the basis for a variational determination of σ^2 in the following sense.

First, we assume for Q and $\delta\tau$ certain 'trial functions' which are arbitrary in the first instance except for the requirement that they satisfy the same boundary conditions as are demanded of the true proper solutions. Then, we evaluate $\delta\psi$ and $\delta\mu$ in terms of the assumed trial functions for Q and $\delta\tau$ with the aid of the initial-value equations (cf. Equations (17) and (18))

$$e^{v}[e^{-v}(\delta\psi + \delta\mu)]_{,\alpha} + \psi_{,\alpha}(\delta\psi - \delta\mu) = \mathfrak{F}_{\alpha} \quad (\alpha = 2, 3), \tag{30}$$

where

$$\mathfrak{F}_{\alpha} = (-1)^{\beta}\left[\frac{e^{2\mu_{\alpha}}Q\omega_{,\beta}}{2\sqrt{-g}} + \delta\tau_{,\alpha} + (2\mu_{\beta} + \psi - v)_{,\alpha}\,\delta\tau\right] \quad (\alpha \neq \beta). \tag{31}$$

Equation (30) represents simple quasi-linear differential equations for $\delta\psi$ and $\delta\mu$ and can be solved by standard methods. Thus

$$(\delta\psi + \delta\mu)_{\text{along } \psi = \text{constant}} = e^{v}\int_{\psi = \text{constant}} e^{-v}\mathfrak{F}_{\alpha}\,dx^{\alpha}, \tag{32}$$

where the integral on the right-hand side is taken along a contour $\psi = $ constant; with $\delta\psi + \delta\mu$ determined by equation (32), $\delta\psi - \delta\mu$ follows directly from Equation (30).

With the assumed trial functions for Q and $\delta\tau$ and the deduced values for $\delta\psi$ and $\delta\mu$, we can formally evaluate σ^2 given by Equation (28). We can similarly evaluate $\sigma^2 + \delta\sigma^2$ which follows from using the trial functions $Q + \delta Q$ and $\delta\tau + \delta^2\tau$ which differ from Q and $\delta\tau$ by some (arbitrarily specified) increments δQ and $\delta^2\tau$. If we now demand that $\delta\sigma^2$ vanishes *identically*, i.e. for *all* δQ and $\delta^2\tau$ (restricted only by the boundary conditions that must be satisfied), then it can be shown that the originally chosen Q and $\delta\tau$ must satisfy the correct dynamical equations of the problem and that σ^2 given by Equation (28) is a true characteristic value.

It is clear from the foregoing remarks that Equation (28) can be used to evaluate the radiation damping of odd-parity, magnetic-type modes of axisymmetric oscillation of the Kerr black hole – 'odd-parity, magnetic-type modes' since the source of the radiation, as is evident from Equation (23), resides in the angular momentum of the field (as manifested by the appearance on the right-hand side of the equation, the function ω which represents the dragging of the inertial frame).

5.1. CARTER'S THEOREM

Equation (28) has an important application to the problem of whether we can deform quasi-stationarily an asymptotically flat axisymmetric vacuum metric external to an event horizon, i.e. to Carter's theorem (1971).

It can be shown that a necessary and sufficient condition for the existence of such a quasi-stationary deformation is obtained by setting both σ^2 and $\delta\tau$ equal to zero in Equation (28). That σ^2 should be set equal to zero for quasi-stationary deformations is clear. That we can also set $\delta\tau = 0$ is a consequence of the fact that during a quasi-stationary deformation, we can continue to maintain the coordinate condition that we are entitled to impose under stationary conditions. Accordingly in the gauge $\mu_2 = \mu_3 = \mu$, Equation (28) gives (see Equation (42) below)

$$\iint \left[e^\beta \, \delta\psi_{,\alpha} \, \delta\psi_{,\alpha} + 2X \, e^{3\psi - \nu}(\delta\psi)^2 - 2(Q_{,2}\omega_{,3} - Q_{,3}\omega_{,2}) \, \delta\psi + \right.$$
$$\left. + \tfrac{1}{4} e^{-3\psi + \nu} Q_{,\alpha} Q_{,\alpha} \right] dx^2 \, dx^3$$
$$= \iint \left[e^\beta \, \delta\psi_{,\alpha} \, \delta\psi_{,\alpha} + 2X \, e^{3\psi - \nu}(\delta\psi)^2 - \tfrac{1}{4} e^{-3\psi + \nu} Q_{,\alpha} Q_{,\alpha} \right] dx^2 \, dx^3 = 0.$$
(33)

The surface integrals in Equation (28) do not survive under the present conditions: at infinity by the requirements of asymptotic flatness and on the horizon and on the axis by the requirements

$$e^\beta = 0 \text{ on the horizon and on the rotation axes;} \tag{34}$$

besides, it must be supposed that all the perturbations vanish on the horizon in such a way that (see below)

$$e^\beta \, \delta\nu \, \delta\nu_{,\alpha} = e^\beta \, \delta\psi \, \delta\psi_{,\alpha} = e^\beta \, QQ_{,\alpha} = Q \, \delta\psi = 0. \tag{35}$$

By a sequence of elementary transformations, the integrands in Equation (32) can be brought to positive-definite forms; and Carter's theorem (that quasi-stationary, axisymmetric deformations, of asymptotically flat axisymmetric vacuum metrics external to an event horizon, are impossible) follows.

There is a direct and a simple way in which we can derive the basic Equation (33) and which clarifies some important aspects of the theorem.

By linearizing the field equations appropriate to the stationary metric (8) about a given solution that corresponds to an axisymmetric black hole, we readily obtain, in the gauge $\mu_2 = \mu_3 = \mu$, the equations

$$(e^\beta \, \delta\beta)_{,\alpha\alpha} = 0, \tag{36}$$

$$(e^{-3\psi+\nu} Q_{,\alpha})_{,\alpha} = [\omega_{,3}(3\delta\psi - \delta\nu)]_{,2} - [\omega_{,2}(3\delta\psi - \delta\nu)]_{,3}, \tag{37}$$

and

$$(e^\beta \, \delta\psi_{,\alpha})_{,\alpha} = 2e^{3\psi-\nu} X \, \delta\psi - (Q_{,2}\omega_{,3} - Q_{,3}\omega_{,2}). \tag{38}$$

From Equation (36) it readily follows that

$$\delta\beta = \delta\psi + \delta\nu = 0; \tag{39}$$

and Equation (37) becomes

$$(e^{-3\psi+\nu} Q_{,\alpha})_{,\alpha} = 4(\omega_{,3} \, \delta\psi_{,2} - \omega_{,2} \, \delta\psi_{,3}). \tag{40}$$

Now multiplying Equations (38) and (40) by $\delta\psi$ and Q, respectively, and integrating over all three-space external to the horizon, we obtain after integrations by parts

$$\iint [e^\beta \, \delta\psi_{,\alpha} \, \delta\psi_{,\alpha} + 2e^{3\psi-\nu} X (\delta\psi)^2 - (Q_{,2}\omega_{,3} - Q_{,3}\omega_{,2}) \, \delta\psi] \, dx^2 \, dx^3 = 0 \tag{41}$$

and

$$\iint e^{-3\psi+\nu} Q_{,\alpha} Q_{,\alpha} \, dx^2 \, dx^3 = 4 \int (Q_{,2}\omega_{,3} - Q_{,3}\omega_{,2}) \, \delta\psi \, dx^2 \, dx^3. \tag{42}$$

The integrated parts vanish in both cases: at infinity by the requirement of asymptotic flatness and on the horizon and on the axes by virtue of the boundary conditions (34) and (35). Equations (41) and (42) are clearly equivalent to the identity from which Carter's theorem follows.

It is important to observe that the proof of Carter's theorem at no stage requires that the perturbations satisfy any continuity requirements on the horizon: all that is needed is that the squares and products of the perturbations vanish on the horizon and also that the derivatives with respect to x^2 and x^3 of such squares and products remain bounded so that when multiplied by e^β they vanish.

5.2. SPHERICALLY SYMMETRIC BLACK HOLES

In the case of spherical symmetry and in the absence of rotation ($\omega = 0$), Equation (28) gives

$$\sigma^2 \iint [e^{-2\nu} \sqrt{-g} \{(\delta\tau)^2 + (\delta\psi)^2 - [\delta(\psi+\mu)]^2\}] \, dx^2 \, dx^3 =$$

$$= \iint [e^\beta [e^{\mu_3-\mu_2}(\delta\psi_{,2})^2 + e^{\mu_2-\mu_3}(\delta\psi_{,3})^2] - 2U(\delta\tau)^2 -$$

$$-2e^{\beta}\{e^{\mu_3-\mu_2}[\beta_{,2}\,\delta\mu_{3,2}-(\psi-v)_{,2}\,\delta\psi_{,2}]-$$
$$-e^{\mu_2-\mu_3}[\beta_{,3}\,\delta\mu_{2,3}-(\psi-v)_{,3}\,\delta\psi_{,3}]\}\,\delta\tau]\,dx^2\,dx^3 + \text{surface integrals},$$
(43)

where for the sake of brevity we have not written out the surface integrals explicitly. It should be noted that the term in Q^2 on the left-hand side of Equation (28) cancels directly with the terms in $(Q_{,2})^2$ and $(Q_{,3})^2$ on the right-hand side by virtue of the equation satisfied by Q.

By using spherical polar coordinates and expanding the various functions in Legendre polynomials, we can separate Equation (43) to obtain the characteristic frequencies belonging to the different harmonics; and this resulting equation can be used to determine the damping of the different non-radial modes of oscillation of the Schwarzschild black hole.

For quasi-stationary deformations, Equation (43) gives, in the gauge $\mu_2 = \mu_3$,

$$\iint e^{\beta}\,\delta\psi_{,\alpha}\,\delta\psi_{,\alpha}\,dx^2\,dx^3 = 0;$$
(44)

and the impossibility of neutral deformations follows. Since the restriction to axisymmetric modes involves no loss of generality when dealing with spherically symmetric systems, Equation (44) excludes general non-axisymmetric deformations as well. (For an alternative demonstration of this result, see Vishveshwara, 1970).

6. On a Criterion for the Occurrence of a Dedekind-Like Point of Bifurcation Along an Axisymmetric Sequence

As we have remarked earlier, radiation-reaction can induce secular instability along a sequence of axisymmetric configurations at a Dedekind-like point of bifurcation. In the framework of general relativity, we can obtain a criterion for the occurrence of such a point of bifurcation by considering the field equations valid for stationary non-axisymmetric systems and linearizing them about a stationary axisymmetric solution for deformations whose dependence on the azimuthal angle is $e^{2i\phi}$.

It is readily seen that for describing stationary non-axisymmetric systems, a suitable form for the metric is (cf. Equation (6))

$$ds^2 = -e^{2\mathfrak{n}}(dt - \mathfrak{w}\,d\phi - q_{2,1}\,dx^2 - q_{3,1}\,dx^3)^2 + e^{2\mathfrak{p}}(d\phi)^2 +$$
$$+ e^{2\mu_2}(dx^2)^2 + e^{2\mu_3}(dx^3)^2,$$
(45)

where $\mathfrak{n}$, $\mathfrak{p}$, $\mathfrak{w}$, q_2, q_3, μ_2, and μ_3 are functions of the three space-variables $\phi\,(=x^1)$, x^2, and x^3 only; and as in the case of the time-dependent axisymmetric metric (6), the three functions $\mathfrak{w}$, q_2, and q_3 can occur in the field equations only in the combinations

$$\mathfrak{w}_{,2} - q_{2,11},\ \mathfrak{w}_{,3} - q_{3,11},\ \text{and}\ q_{2,31} - q_{3,21},$$
(46)

so that we have again only six independent functions to consider.

The metric (46) includes stationary axisymmetric metrics, now written in the form (cf. Equation (8))

$$ds^2 = -e^{2\mathfrak{n}}(dt - \mathfrak{w}\, d\phi)^2 + e^{2\mathfrak{p}}(d\phi)^2 + e^{2\mu_2}(dx^2)^2 + e^{2\mu_3}(dx^3)^2, \qquad (47)$$

where the functions $\mathfrak{n}$, $\mathfrak{p}$, $\mathfrak{w}$, μ_2, and μ_3 are now functions of x^2 and x^3 only. However, as in the case of the metric (8), we have the gauge-freedom to impose a coordinate condition on μ_2 and μ_3.

From a comparison of the metrics (6) and (45), it is clear that we should be able to pass from equations which are valid for time-dependent axisymmetric systems to equations which are valid for time-independent non-axisymmetric systems by some simple rules of transcription. Thus, in place of Equation (12), expressing the conservation of angular momentum per baryon, we now have the conservation equation

$$u^j \left(\frac{\varepsilon + p}{N} u_0 \right)_{,j} = 0. \qquad (48)$$

(It may be noted here that in the Newtonian limit Equation (48) reduces to the Bernoulli integral – a fact one might not have suspected.)

Quite generally, it can be shown that by the replacements

$$u^0 \to -iu^1, \quad u^1 \to iu^0, \quad u_0 \to iu_1, \quad u_1 \to -iu_0,$$

$$\psi \to \mathfrak{n}, \quad \nu \to \mathfrak{p}, \quad \omega \to -\mathfrak{w}, \quad \frac{\partial}{\partial t} \to \Omega \frac{\partial}{\partial \phi},$$

$$V = e^{\psi - \nu}(\Omega - \omega) \to -\frac{1}{V} = -e^{\mathfrak{n} - \mathfrak{p}}\left(\frac{1}{\Omega} - \mathfrak{w}\right),$$

and

$$Q = e^{3\psi - \nu - \mu_2 - \mu_3}(q_{2,3} - q_{3,2}) \to \mathfrak{Q} = e^{3\mathfrak{n} - \mathfrak{p} - \mu_2 - \mu_3}(\mathfrak{q}_{2,3} - \mathfrak{q}_{3,2}), \qquad (49)$$

we can transcribe the equations which are valid for time-dependent axisymmetric systems into equations which are valid for time-independent non-axisymmetric systems.

By subjecting a stationary axisymmetric system (described in conformity with the metric (27)) to non-axisymmetric deformations with a ϕ-dependence

$$e^{im\phi}$$

(where m is an integer greater than or equal to 1), we can write down, with the aid of the rules of transcription (49), the equations that must be satisfied if the system considered can be so deformed quasi-stationarily. And these equations will lead to a characteristic-value problem for m^2 even as the parallel analysis of axisymmetric perturbations with a time-dependence $e^{i\sigma t}$ led to a characteristic-value problem for σ^2. However, in the present context the solution to the characteristic-value problem will be physically meaningful only if the problem allows an integral characteristic value for m. Nevertheless, the characteristic-value problem is *itself* meaningful regardless of whether m happens to be an integer or not. And considered solely as a characteristic-value problem for m^2, we can construct a variational base for evaluating m^2 even as we constructed a variational base in Section 5 for evaluating σ^2. The required formula for m^2, for vacuum metrics, for example, can be written down be letting $\sigma^2 \to -m^2$ and making the replacements (49) in Equation (28). We thus obtain

$$m^2 \iint [2e^{-2p}\sqrt{-g}\{[\delta(\mathfrak{n}+\mu)]^2-(\delta\tau)^2-(\delta\mathfrak{n})^2\} - \tfrac{1}{2}e^{-3\mathfrak{n}-\mathfrak{p}+\mu_2+\mu_3}\mathfrak{Q}^2] \, dx^2 \, dx^3 =$$

$$= \iint [e^{3\mathfrak{n}-\mathfrak{p}} \mathfrak{X}[4(\delta\mathfrak{n})^2+(\delta\tau)^2] + 4e^{3\mathfrak{n}-\mathfrak{p}} \mathfrak{Y} \, \delta\mathfrak{n} \, \delta\tau - 4\mathfrak{U}(\delta\tau)^2 +$$

$$+ 2e^{\beta} [e^{\mu_3-\mu_2}(\delta\mathfrak{n},_2)^2 + e^{\mu_2-\mu_3}(\delta\mathfrak{n},_3)^2] -$$

$$- 4e^{\beta} \{e^{\mu_3-\mu_2}[\beta,_2 \, \delta\mu_{3,2} - (\mathfrak{n}-\mathfrak{p}),_2 \, \delta\mathfrak{n},_2] -$$

$$- e^{\mu_2-\mu_3}[\beta,_3 \, \delta\mu_{2,3} - (\mathfrak{n}-\mathfrak{p}),_3 \, \delta\mathfrak{n},_3]\} \, \delta\tau +$$

$$+ 2[2(\mathfrak{Q},_2 w,_3 - \mathfrak{Q},_3 w,_2) \, \delta\mathfrak{n} - (\mathfrak{Q},_2 w,_3 + \mathfrak{Q},_3 w,_2) \, \delta\tau] +$$

$$+ \tfrac{1}{2}e^{-3\mathfrak{n}+\mathfrak{p}}[e^{\mu_3-\mu_2}(\mathfrak{Q},_2)^2 + e^{\mu_2-\mu_3}(\mathfrak{Q},_3)^2]] \, dx^2 \, dx^3 + \text{surface integrals}$$

where (50)

$$\mathfrak{X} = e^{\mu_3-\mu_2}(w,_2)^2 + e^{\mu_2-\mu_3}(w,_3)^2,$$
$$\mathfrak{Y} = e^{\mu_3-\mu_2}(w,_2)^2 - e^{\mu_2-\mu_3}(w,_3)^2,$$

and

$$\mathfrak{U} = e^{\beta}[e^{\mu_3-\mu_2}(\beta,_2\mu_{3,2}+\mathfrak{n},_2\mathfrak{p},_2) + e^{\mu_2-\mu_3}(\beta,_3\mu_{2,3}+\mathfrak{n},_3\mathfrak{p},_3)]. \quad (51)$$

In Equation (50), the functions that are to be varied are $\mathfrak{Q}$ and $\delta\tau$ while $\delta\mathfrak{n}$ and $\delta\mu$ are to be evaluated in terms of them with the aid of the initial-value equations (cf. Equations (30) and (31))

$$e^{\mathfrak{p}}[e^{-\mathfrak{p}}(\delta\mathfrak{n}+\delta\mu)],_\alpha + \mathfrak{n},_\alpha(\delta\mathfrak{n}-\delta\mu) = \mathfrak{F}_\alpha, \quad (\alpha=2,3) \quad (52)$$

where

$$\mathfrak{F}_\alpha = (-1)^\alpha e^{2\mu_\alpha} \frac{\mathfrak{Q}w,_\beta}{2\sqrt{-g}} + (-1)^\beta [\delta\tau,_\alpha + (2\mu_\beta+\mathfrak{n}-\mathfrak{p}),_\alpha] \quad (\alpha \neq \beta). \quad (53)$$

The variational expression for m^2, including the terms in the fluid variables, has been written down by Chandrasekhar and Friedman (1973c); it can be used to determine whether Dedekind-like points of bifurcation occur along given equilibrium sequences of axisymmetric configurations. If such points of bifurcation occur, then we may anticipate the onset of secular instability by radiation-reaction at these points.

Returning to Equation (50), we shall consider its application to the Kerr metric with a view to determining whether along the Kerr sequence a Dedekind-like point of bifurcation occurs.

By writing the Kerr metric in the Boyer-Lundquist coordinates in the form (47), we find that

$$e^{2\nu} = \frac{\Sigma}{\varrho^2}, \quad e^{2\psi} = \frac{\Delta\varrho^2 \sin^2\theta}{\Sigma}, \quad e^{2\beta} = \Delta \sin^2\theta,$$

$$e^{2\mu_2} = \frac{\varrho^2}{\Delta}, \quad e^{2\mu_3} = \varrho^2, \quad \text{and} \quad \mathfrak{w} = -\frac{2Mar\sin^2\theta}{\Sigma}, \tag{54}$$

where

$$\Delta = r^2 - 2Mr + a^2, \quad \Sigma = r^2 - 2Mr + a^2\cos^2\theta, \quad \text{and} \quad \varrho^2 = r^2 + a^2\cos^2\theta. \tag{55}$$

It should be noted that

$$\Delta = 0 \text{ on the horizon and } \Sigma = 0 \text{ on the stationary limit.} \tag{56}$$

Inserting the foregoing expressions in Equation (50), we obtain

$$m^2 \iint \left[\frac{2\Sigma}{\Delta\sin\theta} [(\delta n + \delta\mu)^2 - (\delta\tau)^2 - (\delta n)^2] - \frac{\varrho^4}{2\Sigma\Delta\sin\theta} \mathfrak{Q}^2 \right] dx^2\, dx^3 =$$

$$= \iint \left\{ \frac{4}{\Sigma^2} \left[\frac{Ma\sin^2\theta}{\varrho^2} (r^2 - a^2\cos^2\theta)(2\delta n + \delta\tau) - \tfrac{1}{2}\varrho^2\mathfrak{Q}_{,3} \right]^2 + \right.$$

$$+ \frac{16\Delta}{\Sigma^2} \left[\frac{Mar\sin\theta\cos\theta}{\varrho^2} (2\delta n - \delta\tau) - \tfrac{1}{4}\varrho^2\mathfrak{Q}_{,2} \right]^2 -$$

$$- \frac{\varrho^4\sin\theta}{2\Sigma^2} [\Delta(\mathfrak{Q}_{,2})^2 + (\mathfrak{Q}_{,3})^2] +$$

$$+ 4\Delta\sin\theta \left[\frac{M}{\Sigma\varrho^2}(r^2 - a^2\cos^2\theta)\delta\tau + \delta n_{,2} \right]^2 +$$

$$+ 16\sin\theta \left[\frac{Ma^2r\sin\theta\cos\theta}{\Sigma\varrho^2}\delta\tau + \tfrac{1}{2}\delta n_{,3} \right]^2 -$$

$$- 2\sin\theta[\Delta(\delta n_{,2})^2 + (\delta n_{,3})^2] - 4\sin\theta \frac{(r-M)^2 - a^2\cos^2\theta}{\Sigma}(\delta\tau)^2 -$$

$$\left. - 4\sin\theta[(r-M)(\delta\mu_{,2} + \delta n_{,2}) - \cot\theta(\delta\mu_{,3} + \delta n_{,3})]\delta\tau \right\} dx^2\, dx^3 +$$

$$+ \text{surface integrals.} \tag{57}$$

We have not explicitly written out the expressions for the surface integrals in Equation (57) since the extent of their survival depends on the boundary conditions which obtain on the horizon and at infinity. (There is, however, no difficulty in showing that we have no contributions from infinity if the requirements of asymptotic flatness are met.)

We observe that as written out, the integrands on both sides of Equation (57) appear to diverge on the stationary limit at $\Sigma=0$ *and* on the horizon at $\Delta=0$. However, by examining in detail the equations governing the perturbations, we find that on the stationary limit, the equations allow solutions with the behaviours

$$\delta\mathfrak{n} = n\Sigma^2, \quad \delta\mathfrak{p} = p\Sigma^2, \quad \delta\tau = t\Sigma^2, \quad \delta\mu = m\Sigma^2, \quad \text{and} \quad \mathfrak{Q} = \frac{\phi}{\varrho^2}\Sigma^3, \tag{58}$$

where

$$n = \tfrac{1}{4}\phi\left(\frac{r-M}{a\cos\theta} - \frac{a\cos\theta}{r-M}\right) = -p,$$

$$m = -\tfrac{1}{4}\phi\left(\frac{r-M}{a\cos\theta} - \frac{a\cos\theta}{r-M}\right),$$

$$t = -\tfrac{1}{12}\phi\left(\frac{r-M}{a\cos\theta} + \frac{a\cos\theta}{r-M}\right), \tag{59}$$

and ϕ is an arbitrary function on $\Sigma=0$. Similarly, we find that on the horizon, the equations allow solutions with the behaviours

$$\mathfrak{Q} = -\frac{4CMar\cos\theta}{\varrho^4(M^2-a^2)^{1/2}}\Delta^{3/2},$$

$$\delta\tau = C\Delta^{1/2}, \quad \delta\mu = -C\Delta^{1/2},$$

$$\delta\mathfrak{p} = C\left(\frac{m^2 a^2}{M^2-a^2} - 1\right)\Delta^{1/2},$$

and

$$\delta\mathfrak{n} = \frac{CM(r^2 - a^2\cos^2\theta)}{\varrho^2 a^2(M^2-a^2)^{1/2}\sin^2\theta}\Delta^{3/2}, \tag{60}$$

where C is a constant.

With the foregoing behaviours on $\Sigma=0$ and $\Delta=0$, the integrands in Equation (50) are bounded and the integrals are well defined. But it may be argued that the behaviours on the horizon given by Equations (60) are 'unacceptable' since the derivatives of these functions are singular here. It is, however, worth noting that the behaviours we find satisfy the minimal requirements which we found in Section 5.1 were necessary for the validity of Carter's theorem. And apart from the acceptability or otherwise of the behaviours given in Equations (60), the question remains whether a value for m (not necessarily integral) exists for which the perturbation equations allow solutions with the behaviours (58) and (60) on the stationary limit and on the horizon (respectively) and still satisfy the requirements of asymptotic flatness at infinity. If such solutions do exist, the next question concerns whether along the Kerr sequence there is a point where m has the value 2. And even if such a point exists, the question whether the occurrence of such a point signifies the secular instability of the Kerr metric at that point will still remain.

It is realized that the foregoing remarks are at variance with the conclusions that have been reached by Press and Teukolsky (1973), Wald (1973), and Stewart (1973). But it is not clear to the writer whether the regularity requirements imposed by these authors on the horizon are not too severe. In any event it would appear that one should be able to arrive at the correct result by the present analysis equally well.

Acknowledgement

The research reported in this paper has in part been supported by the National Science Foundation under grant GP-34721X1 with the University of Chicago.

References

Bardeen, J. M., Thorne, K. S., and Meltzer, David W.: 1966, *Astrophys. J.* **145**, 505.
Carter, Brandon: 1971, *Phys. Rev. Letters* **26**, 331.
Chandrasekhar, S.: 1964, *Astrophys. J.* **140**, 417.
Chandrasekhar, S.: 1969, *Ellipsoidal Figures of Equilibrium*, Yale University Press, New Haven.
Chandrasekhar, S.: 1970, *Astrophys. J.* **161**, 561.
Chandrasekhar, S.: 1974, *Astrophys. J.* **187**, 169.
Chandrasekhar, S. and Friedman, J.: 1972a, *Astrophys. J.* **175**, 379.
Chandrasekhar, S. and Friedman, J.: 1972b, *Astrophys. J.* **176**, 745.
Chandrasekhar, S. and Friedman, J.: 1973a, *Astrophys. J.* **177**, 745.
Chandrasekhar, S. and Friedman, J.: 1973b, *Astrophys. J.* **181**, 481.
Chandrasekhar, S. and Friedman, J.: 1973c, *Astrophys. J.* **185**, 1.
Chandrasekhar, S. and Lebovitz, N.: 1973, *Astrophys. J.* **185**, 19.
Durisen, Richard H.: 1973a, *Astrophys. J.* **183**, 205.
Durisen, Richard H.: 1973b, *Astrophys. J.* **183**, 215.
Fowler, W. A.: 1964, *Rev. Mod. Phys.* **36**, 545.
Hartle, J. B. and Thorne, K. S.: 1968, *Astrophys. J.* **153**, 807.
Ipser, J. R. and Detweiler, S.: 1973, *Astrophys. J.*, in press.
Lebovitz, N.: 1970, *Astrophys. J.* **160**, 701.
Ostriker, J. P. and Bodenheimer, P.: 1973, *Astrophys. J.* **180**, 171.
Ostriker, J. P. and Tassoul, J. L.: 1969, *Astrophys. J.* **155**, 987.
Press, W. A. and Teukolsky, S. A.: 1973, in press.
Stewart, John: 1973, in press.
Tassoul, J. L. and Ostriker, J. P.: 1968, *Astrophys. J.* **154**, 613.
Teukolsky, S. A.: 1972, *Phys. Rev. Letters* **29**, 1114.
Teukolsky, S. A.: 1973, in press.
Thorne, K. S.: 1969, *Astrophys. J.* **158**, 1.
Thorne, K. S.: 1969, *Astrophys. J.* **158**, 997.
Thorne, K. S. and Campolattaro, A.: 1967, *Astrophys. J.* **149**, 591.
Thorne, K. S. and Campolattaro, A.: 1970, *Astrophys. J.* **159**, 847.
Thorne, K. S. and Meltzer, D. W.: 1966, *Astrophys. J.* **145**, 514.
Thorne, K. S. and Price, Richard: 1969, *Astrophys. J.* **155**, 163.
Vishveshwara, C. V.: 1970, *Phys. Rev.* **D1**, 2870.
Wald, Robert M.: 1973, in press.

GRAVITATIONAL COLLAPSE

R. PENROSE

University of Oxford, England

Abstract. In the standard picture of gravitational collapse to a black hole, a key role is played by the hypothesis of cosmic censorship – according to which no naked space-time singularities can result from any collapse. A precise definition of a naked singularity is given here which leads to a strong 'local' version of the cosmic censorship hypothesis. This is equivalent to the proposition that a Cauchy hypersurface exits for the space-time. The principle that the surface area of a black hole can never decrease with time is presented in a new and simplified form which generalizes the earlier statements. A discussion of the relevance of recent work to the naked singularity problem is also given.

The theoretical picture of gravitational collapse to a black hole is now a familiar one (Penrose, 1969; Hawking and Ellis, 1973), so I shall dwell only briefly upon it. The essentials are depicted in Figure 1. Matter collapses inwards under the influence of gravitation until a situation is reached from which there is no escape. We may rec-

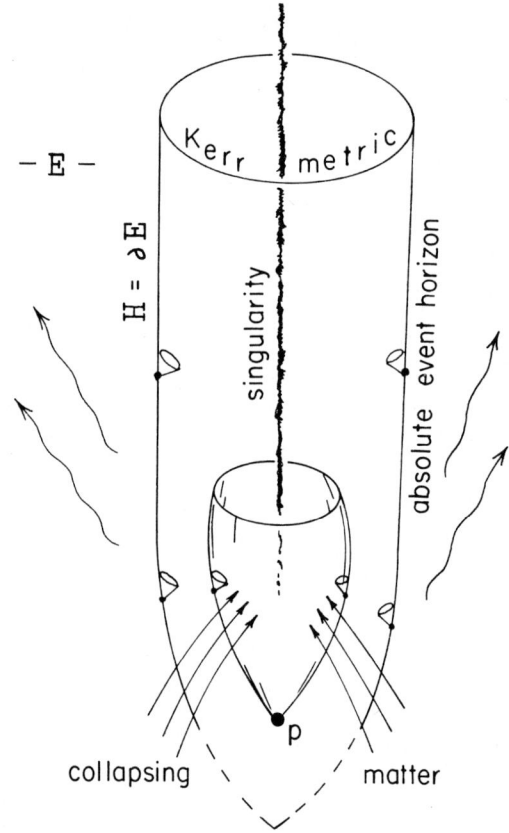

Fig. 1. Gravitational collapse to a black hole according to a standard viewpoint: the future light cone of p starts to reconverge and the external field settles down to that of a Kerr black hole.

ognise that such a situation has occurred by the presence of a *trapped surface* (Penrose, 1968) or, as in the case depicted, by the fact that sufficient matter crosses the future light cone of some event p to cause the divergence of the light rays (null geodesics through p) to change sign, so that the rays begin to converge towards one another again. When this occurs we can invoke the Hawking-Penrose theorem (Hawking and Penrose, 1970) to deduce the existence of some form of *space-time singularity*. To derive this conclusion we need only assume, in addition to Einstein's equations, that the density plus each principal pressure is non-negative and the density plus the sum of the principal pressures is likewise non-negative, that there are no closed timelike curves and that the space-time is appropriately 'general' – all of which are very reasonable requirements from the physical point of view.

At this juncture the assumption of *cosmic censorship* can be brought in (Penrose, 1969, 1974a). This hypothesis forbids the existence of a *naked singularity*, i.e. any space-time singularity arising in a collapse from which a signal can originate which escapes to infinity. I shall discuss the status of this assumption a little later. For the moment I merely point out that it is the most *conservative* of the positions available to us. The violation of cosmic censorship would lead to a picture of gravitational collapse much more radical than that of a black hole.

Consider, now, the set E consisting of all events from which a timelike curve (or a null curve – it makes little difference) can be drawn into the future to infinity. The boundary ∂E of E is called the *absolute event horizon* (Penrose, 1969). Thus the absolute event horizon may be thought of as the boundary between the regions consisting of those events from which an observer can escape to infinity and those events from which he cannot. The principle of cosmic censorship may now be adopted in the form which states that all the resulting space-time singularities must be surrounded by – and must not lie on – the absolute event horizon $H = \partial E$. One precise statement of this hypothesis is Hawking's condition of future asymptotic predictability (Hawking, 1972; Hawking and Ellis, 1973). This applies, strictly speaking, only in the case of an asymptotically flat space-time. I shall give a more general criterion shortly. When such a non-singular absolute event horizon exists we say that the universe model contains a black hole or black holes – depending upon the number of disconnected pieces into which this horizon falls. We may also envisage two or more black holes congealing into one. Then the absolute event horizon resembles the situation depicted in Figure 2. When a black hole settles down into a stationary state then, as is virtually established by the results of Israel (1967), Carter (1971), Hawking (1972), and others (Müller zum Hagen et al., 1972; cf. also Hawking and Ellis, 1973), it takes up the configuration of a *Kerr* metric. This is defined by just two parameters m and a, where m specifies the mass and am the angular momentum. For a black hole, $m \geq a$.

An important feature of black holes subject to cosmic censorship is the area principle. This states that if S_1 and S_2 are two cross-sections of H, where S_2 lies entirely to the future of S_1 along H, then the area of S_1 cannot exceed that of S_2. (Each S_i is a spacelike 2-surface, not necessarily smooth). Hawking (1972) has used the area

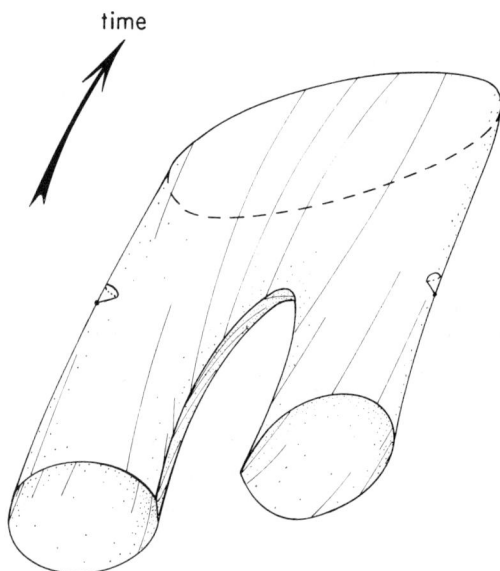

Fig. 2. Two black holes congealing to form a single black hole.

principle to derive an uuper bound to the amount of gravitational radiation emitted when two black holes coalesce. For this, the area principle is used in the form

$$A_1 + A_2 \leqslant A_3,$$

where the subscripts 1 and 2 refer to the colliding black holes and 3 to the resultant black hole, the surface area A_i of each black hole being given by

$$A_i = 8\pi m_i (m_i + (m_i^2 - a_i^2)^{1/2}),$$

when the black holes are settled into their respective Kerr configurations. Thus, in this case the cross-section initially consists of two pieces and finally of one piece, but the area principle still holds.

It is worth indicating here the basic properties which lie behind the area principle. The key fact is that every point of H is the past end-point of some future-endless null geodesic lying on H, of infinite affine extent. If δA is an element of surface area of some spacelike cross-section of H and if we propagate δA along the null geodesic generators of H, we get the equation (cf. Penrose, 1968)

$$-\frac{d^2 (\delta A)^{1/2}}{dv^2} = (\delta A)^{1/2} \{\sigma \bar{\sigma} + \Phi_{00}\} \geqslant 0,$$

where v is an affine parameter along the relevant null geodesic γ, where σ is its complex shear, and Φ_{00} is minus one-half the Ricci tensor component associated with γ. The *weak energy condition* (energy plus each principal pressure non-negative) gives $\Phi_{00} \geqslant 0$. It follows that if $d(\delta A)^{1/2}/dv$ ever becomes negative, then $(\delta A)^{1/2}$ must decrease to zero within a finite v value. But if this were to happen, then γ could not remain on H for values of v beyond this (H being the boundary of a set, namely E, which is its own past, cf. Penrose, 1972). This would contradict the statement that each null geodesic generator of H extends into the future to arbitrarily large affine parameter values. Thus $d(\delta A)^{1/2}/dv \geqslant 0$, whence $d(\delta A)/dv \geqslant 0$ along γ.

This shows that the surface area increases along γ as we proceed into the future. Furthermore γ never leaves H in *future* directions. But it can leave H in *past* directions. Thus, the surface area of cross-sections of H can increase into the future for two reasons, either because it increases along any *given* γ or because *new* γ's emerge. (Note that in the black hole collision of Figure 2, both of these situations occur.) Neither of these situations can occur in past directions. Thus, the area principle is established provided it can be shown that null geodesics generate H, having infinite affine lengths into the future. In fact, it is only the condition of infinite affine length which causes any difficulty. It is this that requires a suitable statement of the cosmic censorship principle.

In order to state this principle precisely we require a definition of a space-time singularity. I shall adopt a definition which is not quite the same as that suggested by the singularity theorems, but it is more useful for present purposes. The ideas are taken from Geroch *et al.* (1972) and are somewhat similar to those of Seifert (1971). Let us assume, for simplicity, that the space-time is strongly causal (cf. Hawking and Ellis, 1973; Penrose, 1972). We adjoin some extra 'ideal points' to the space-time which are end-points to future-endless or to past-endless causal (i.e. timelike or null) curves. We say that two future-endless causal curves γ, γ' have the same ideal future end-point if the two curves have identical pasts (i.e. $I^-[\gamma] = I^-[\gamma']$ in the notation of Geroch *et al.* (1972). Such sets are called terminal indecomposable past sets, or TIP's for short). Similarly, two past-endless causal curves have the same ideal past end-point if they have identical futures (called TIF's). (To be able to say when a future-endless causal curve has the same future end-point as the past end-point of some past-endless causal curve would require some extra complications which I do not propose to enter into here.) These ideal points form a sort of boundary to the space-time. But the boundary points need not be singularities. They may be points at infinity. To distinguish these two possibilities we may adopt one of a number of slightly different criterea. Let us choose the simplest one here and say that the boundary point is at infinity if and only if there is a semi-infinite causal curve of infinite proper length which has that boundary point as an ideal end-point. Otherwise, we can say we have a finite boundary point. In the case that we have a space-time which is maximally extended (i.e. not a proper part of any other connected space-time), then all *finite* boundary points may be reasonably interpreted as *singular* points of the space-time. One might also consider some of the points at infinity also to be singular, but I shall not bother with this possibility here.

Let us now try to interpret the cosmic censorship hypothesis in these terms. We need a definition of a naked singularity, but we must be careful not to rule out the 'big bang' in the exclusion of naked singularities, even though the big bang singularities of the normal cosmological models are indeed naked in the sense that signals can escape from them (i.e. they are represented, in the above descriptions by past end-points of past-endless causal curves). The idea of a naked singularity is that it might arise in the collapse of ordinary matter from a non-singular initial state. I shall formalize this idea by regarding a singularity as naked if there is some observer (timelike curve) for whom the singularity lies initially to his future and subsequently to his past. This concept is in essence time-symmetrical, and it should be observed that the normal cosmological big bang does not qualify as a naked singularity in this sense. The definition is basically 'local' in that no mention is made of signals escaping to infinity. Thus, even a singularity inside a black hole might conceivably be 'naked' to some observer who is himself inside the black hole. However, in the normal picture of spherically symmetrical collapse, such a situation does not occur. There is some indication also (Simpson and Penrose, 1973) that in a generic perturbed collapse this situation still does not occur.

Now we can apply the above idea to the aforementioned definition of a singularity either of two ways around. Let us suppose, first, that we are concerned with the ideal *past* end-point q of a past-endless timelike curve γ (i.e. with the TIF $I^+[\gamma]$). We can say that q lies to the (causal) future of a point p if the future of p contains the future of γ (i.e. $I^+(p) \supset I^+[\gamma]$). In this case, any point r of γ lies to the future of p, so there are timelike curves from p to r. An observer following one of these timelike curves will have q to his future when he is at p and to his past when he is at r. (See Figure 3.)

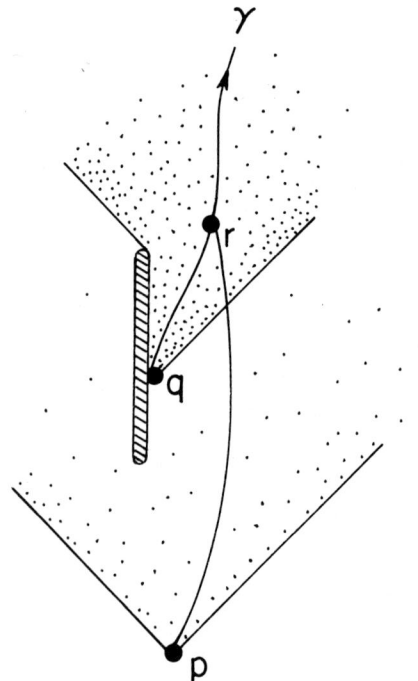

Fig. 3. A naked singularity q – the future of some point p contains the future of q.

Thus, if q is a singular point, then this is the situation where a naked singularity occurs. A similar situation might possibly arise when q is a point at infinity. For purposes of the present discussion I want to regard such a possibility as something akin to a naked singularity. The normal cosmological models do not admit such an occurrence. (The complete anti-de Sitter space is an exception, cf. Penrose (1968), but cannot be regarded as a 'normal' cosmological model.) Also, while this possibility *can* occur in certain idealized models of collapse, there is some indication that it may be unstable in the sense that small perturbations could lead to genuine curvature singularities (cf. Simpson and Penrose, 1973). I shall refer to the hypothesis which excludes the above situation from occurring whether or not q is a point at infinity as *strong cosmic censorship*.

We could also apply the above argument in a time-reversed sense, allowing q to be the future end-point of a future-endless timelike curve γ'. It turns out (although this is not immediately obvious) that we get precisely the same condition of strong cosmic censorship as above if we proceed in this time-reversed way, i.e. if we rule out any point r whose past could contain the past of such a curve γ'. In fact, it can be shown (cf. Penrose, 1972) that strong cosmic censorship is equivalent to Leray's condition of *global hyperbolicity*. Using a theorem of Geroch (1970) we can state global hyperbolicity to be the condition that a Cauchy hypersurface exists for the space-time. Geroch's theorem also implies that the entire universe model is then topologically the product of this Cauchy hypersurface with an open timelike line. Thus, we may say that if the strong cosmic censorship principle holds, then the topology of the universe is unchanging for all time. (I shall discuss this more fully, in the case of a closed universe model (Penrose, 1974b)). In fact, global hyperbolicity leads to many other simplifications in the global structure of a space-time. For example, any two points with a timelike separation in such a space-time can be connected by a timelike geodesic which maximizes the lengths of all timelike curves which connect the points. Thus, if strong cosmic censorship turns out to be a reasonable requirement for realistic space-time models, then many global problems connected with cosmology and black-hole structure can be handled without fear of the sort of pathology arising which could complicate so much of the earlier global analysis.

As one example of the utility of strong cosmic censorship, let me give a statement of the area principle which generalizes the particular version for asymptotically flat space-times given by Hawking (1972) (cf. also Hawking and Ellis, 1973). Suppose the space-time is globally hyperbolic and the energy condition holds everywhere. Define E to be the set of points which are past end-points of timelike curves of infinite length. The required result is that the *absolute event horizon* ∂E then satisfies the area principle (i.e. if S_1 and S_2 are two sections of ∂E, with S_2 to the future of S_1, then the area of S_2 is not less than that of S_1). To specialize this result to the situation considered by Hawking, namely to a space-time which is future asymptotically predictable from some spacelike hypersurface K, we need consider, for our space-time manifold, only that portion of the entire space-time which is the interior of the domain of dependence of K (cf. Hawking and Ellis, 1973; Penrose, 1972). But the generalized version of the result as stated here may be applied also to cosmological situations. I shall give a proof and discussion of the result elsewhere.

What is the theoretical evidence in favour of or against cosmic censorship (not necessarily in its strong form)? In my opinion the evidence in either direction is very scanty indeed. I think that in a collapse which does not differ very greatly in its initial conditions from the standard spherically symmetrical collapse situation I described first, then some form of cosmic censorship is likely to be valid. The analysis by various workers of the stability of the Kerr metric tends to confirm this belief. (Much of this work is being described elsewhere in this symposium (cf. Chandrasekhar, 1974; Press, 1974; Teukolsky, 1974).) It seems likely that this programme will come to a successful conclusion in the near future and that the Kerr solution (for all m, a with $m > a$) will be declared 'officially' to be stable. Perhaps I can be forgiven, on the other hand, if I pose a few queries concerning the whole question of stability for a black hole. I am not trying to cast any doubt on the existing analysis but merely suggest some further questions that one might attempt to answer.

In the first place, it is not completely clear to me that it is legitimate to assume, in the context of establishing that all solutions settle into a Kerr configuration, that a black hole is ever exactly stationary. I have in mind the situation which arose in connection with the Newmann-Penrose constants for asymptotically flat space-times. In fact, one can show quite rigorously that a transition from one exactly stationary state to another exactly stationary state via an intermediate phase in which gravitational radiation is emitted cannot occur unless a certain combination of multipole moments returns to its original value (Newman and Penrose, 1968). But this does *not* mean that this multipole moment combination must return to its original value in order for the system to settle down. It is simply that the presence of a small and ever-decreasing disturbance in the distant gravitational field (the backscattered gravitational radiation) is always just sufficient to prevent any inference concerning the final multipole moment combination to be drawn. I think it is unlikely that anything of this nature could arise to spoil the black hole results, but the issue is not clear to my mind.

It is also not quite clear to me whether the perturbation analysis is actually aimed at excluding the possibility that the absolute event horizon might itself develop into a curvature singularity – which would then be a naked singularity. A 'perturbation' which is actually singular on the event horizon might be excluded on the basis that it represents unreasonable initial data with which to start, but this becomes less clear if the black hole is never assumed to have quite reached a stationary state.

These points are perhaps subtleties and may well have no real significance for the question of black hole stability. However, a more serious question concerns the stability of a black hole near the critical case $a = m$. If a is only marginally smaller than m and a perturbation is applied which is in some sense comparable with the difference $m - a$, then it is by no means clear that such a perturbation can be regarded as small. It is difficult to see how this sort of disturbance could be analyzed within the framework of perturbation theory.

Of course, the whole question of large disturbances applied to black holes is quite an open one. For example, it is often assumed that if two black holes of comparable mass are brought together, then the result will again be a black hole. While I would agree that this is certainly one clear possibility, it is by no means obvious that it is the only one. The possibility that two black holes might collide to form a naked singularity is excluded *only* by the pure assumption that the cosmic censorship hypothesis holds. This would again be a question of an absolute event horizon developing into a naked singularity – a circumstance which I do not see how to exclude on theoretical grounds. Perhaps the computer analysis by DeWitt and his co-workers of a black hole head-on collision will shed some light on this question. It is also possible that two black holes spiralling into one another might produce a qualitatively different result from a head-on collision.

In view of the above uncertainties it is worthwhile to investigate whether or not it is possible to set up initial states of collapse for which a contradiction with the standard picture might be obtained. The combination of area principle, mass-energy conservation, and the final situation of a Kerr metric, together impose definite constraints on the initial geometry. Basically, one is not allowed an initial situation involving a trapped surface of too great area for the initial mass involved. I have described elsewhere (Penrose, 1974a) an attempt at obtaining a contradiction with the standard picture (and hence with cosmic censorship) in this way. Some partial results by Gibbons (1973) have made it seem unlikely that a contradiction with the standard picture can be arrived at by such considerations alone.

A more direct attack on the naked singularity question is that of Müller zum Hagen *et al.* (1973). They construct an explicit solution of Einstein's equations which describes a collapsing dust cloud which eventually forms a black hole. However, before doing so, the dust encounters caustics at which the density becomes infinite. For a short while this region of infinite density – and therefore infinite curvature – is visible from infinity (Figure 4). Thus the solution must be said genuinely to describe a collapse with a naked singularity. It seems that the solution should be stable under small perturbations of the initial state since Grischuk (1967) has shown that fully general timelike singularities (involving rotation) can occur with dust. Müller zum Hagen *et al.* (1973) also show that in the spherically symmetrical case, the introduction of a bounded pressure does not substantially affect their conclusions: a naked singularity can still arise. Stability of the singularity situation for such a fluid under general perturbations (which involve some rotation) is not considered however.

I think that these examples are interesting more for the questions that they raise than for the questions that they answer. I certainly do not feel that in themselves they overthrow the cosmic censorships principle, but they do cause one to wonder what form of precise statement such a principle should have, if one is to have any hope of proving it. Naked singularities which are not stable under perturbations of the initial conditions should presumably be discounted. But what about perturbations of the equations of state? What kinds of such perturbation should be permitted? Or would it be simpler to restrict attention just to the vacuum case? What role should be played by considerations such as those of Hagedorn (1968) according to which the maximum permissible pressure-to-density ratio goes down to zero as the density increases to infinity? If this is valid, the material would behave more and more like dust as the density increases.

Finally, we should ask what is the observational status of naked singularities.

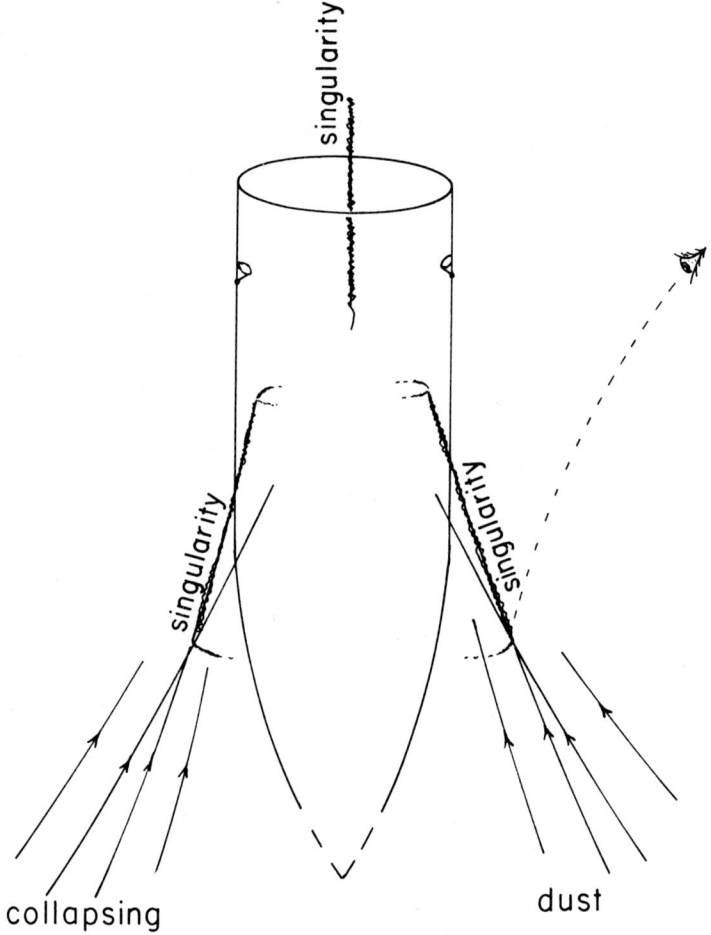

Fig. 4. The Müller zum Hagen-Seifert-Yodzis model of a collapsing dust cloud with naked singularity of the 'shell-crossing' type.

There are certainly many strange phenomena observed to occur, especially in galactic nuclei and in quasars. The temptation to invoke some sort of naked singularity as an explanation must be strong. But the reasons for doing so would seem to be of a rather negative character. The physics at a naked singularity would be largely unknown, so explanations of this kind would have little predictive power. I feel that there is one possible exception, however, and this is in Weber's observations. If his results turn out to be substantiated, then we shall be faced with a problem of energy balance which seems to have little hope of solution unless a very effective beaming mechanism can be suggested which sends the gravitational waves out closely in the plane of the galaxy. Misner's ingenious suggestion (cf. Misner, 1974) for doing this by means of gravitational synchrotron radiation from particles orbiting a black hole seems now not to be feasible. There is an alternative possibility whereby a naked Kerr ring singularity achieves such beaming (Penrose, 1974a) and this should also be considered. But only if the observations pointed clearly to the necessity of such an explanation could one be expected to take such suggestions very seriously.

References

Carter, B.: 1971, *Phys. Rev. Letters* **26**, 331.
Chandrasekhar, S.: 1974, this volume, p. 63.
Geroch, R.: 1970, *J. Math. Phys.* **11**, 437.
Geroch, R., Kronheimer, E. H., and Penrose, R.: 1972, *Proc. Roy. Soc. London* **A327**, 545.
Gibbons, G.: 1973, to appear.
Grischuk, L. P.: 1967, *Sov. Phys. J.E.T.P.* **24**, 320.
Hagerdorn, R.: 1968, *Nuovo Cimento* **56A**, 1027.
Hawking, S. W.: 1972, *Comm. Math. Phys.* **25**, 152.
Hawking, S. W. and Ellis, G. F. R.: 1973, *The Large Scale Structure of Space-Time*, Cambridge Univ. Press.
Hawking, S. W. and Penrose, R.: 1970, *Proc. Roy. Soc. London* **A314**, 529.
Israel, W.: 1967, *Phys. Rev.* **164**, 1776.
Misner, C. W.: 1974, this volume, p. 3.
Müller zum Hagen, H., Robinson, D. C., and Seifert, H.-J.: 1972, *Gen. Relativity and Gravitation* **4**, 1, 53.
Müller zum Hagen, H., Seifert, H.-J., and Yodzis, P.: 1973, *Comm. Math. Phys.* **34**, 135.
Newman, E. T. and Penrose, R.: 1968, *Proc. Roy. Soc. London* **A305**, 175.
Penrose, R.: 1968, in C. M. De Witt and J. A. Wheeler (eds.), *Battelle-Rencontres*, Benjamin, New York.
Penrose, R.: 1969, *Rivista del Nuovo Cimento, Ser. 1*, **1**, Num. Spec. 252.
Penrose, R.: 1972, *Techniques of Differential Topology in Relativity*, S.I.A.M., Philadelphia.
Penrose, R.: 1974a, *Ann., N.Y. Acad. Sci.* **224**, 125.
Penrose, R.: 1974b, in M. S. Longair (ed.), 'Confrontation of Cosmological Theories with Observational Data', *IAU Symp.* **63**, in press.
Press, W. H.: 1974, this volume, p. 93.
Seifert, H.-J.: 1971, *Gen. Relativity Gravitation* **1**, 247.
Simpson, M. and Penrose, R.: 1973, *Int. J. Theor. Phys.* **7**, 183.
Teukolsky, S.: 1974, this volume, p. 92.

PERTURBATIONS OF A ROTATING BLACK HOLE

SAUL A. TEUKOLSKY
California Institute of Technology, Pasadena, Calif., U.S.A.

Abstract. Decoupled, separable wave equations describing neutrino, electromagnetic and gravitational perturbations have been derived (Teukolsky, 1972, 1973). A number of applications have been made (Press and Teukolsky, 1973; Starobinsky and Churilov, 1973).

References

Press, W. H. and Teukolsky, S. A.: 1973, *Astrophys. J.* **185**, 649.
Starobinsky, A. A. and Churilov, S. M.: 1973, *Zh. E.T.F.* **65**, 3.
Teukolsky, S. A.: 1972, *Phys. Rev. Letters* **29**, 1114.
Teukolsky, S. A.: 1973, *Astrophys. J.* **185**, 635.

RECENT WORK ON KERR STABILITY AND SUPERRADIANT WAVE SCATTERING

WILLIAM H. PRESS

California Institute of Technology, Pasadena, Calif., U.S.A.

Abstract. As a stopgap measure until a rigorous analytic determination is made, Teukolsky and I have tested the dynamical stability of the Kerr metric under small perturbations numerically (Press and Teukolsky, 1973). We find that it is stable for all $a \leqslant M$. We have also computed the magnitude of electromagnetic and gravitational-wave amplification in superradiant scattering, and Bardeen has independently obtained identical results (Teukolsky, 1973; Teukolsky et al., 1974). The amplification ranges up to $\sim 2\%$ for electromagnetism ($l=m=1$) and up to 140% for gravitation ($l=m=2$); these values are also consistent with Starobinsky's (1973) results for the value of critical frequencies.

References

Press, W. H. and Teukolsky, S. A.: 1973, *Astrophys. J.* **185**, 649.
Starobinsky, A. A.: 1973, *Zh.E.T.F.* **65**, 3.
Teukolsky, S. A.: 1973, unpublished Ph.D. Thesis, available from University Microfilms, Inc., Ann Arbor, Michigan.
Teukolsky, S. A., Bardeen, J. M., and Press, W.: 1974, in preparation.

AMPLIFICATION OF WAVES REFLECTED FROM KERR BLACK HOLES

A. A. STAROBINSKY

The Landau Institute for Theoretical Physics, Moscow., U.S.S.R.

Abstract. The effect of amplification of electromagnetic and gravitational waves reflected from a rotating black hole ('superradiance scattering') is investigated. This effect was proposed by Zel'dovich (1971). It leads, as well as the Penrose process, to the energy extraction from a Kerr black hole at the expense of its rotational energy and momentum decrease. The coefficient of wave reflection $R > 1$ if $\omega < n\Omega$, where ω is the wave frequency, n – its angular momentum and Ω is the black hole angular velocity. The value of this effect is not small in the case of gravitational waves, for example, if $l = n = 2$, $\omega \to n\Omega$ and $a = M$, then $R \approx 2.38$.

There also exists a quantum version of the effect, namely, the one of spontaneous pair creation in the Kerr metric, but this quantum effect is exceedingly small in real astrophysical conditions, because its characteristic time is of the order $G^2 M^3 / \hbar c^4$, where M is the black hole mass.

References

Zel'dovich, Ya. B.: 1971, *Pisma v Zh.E.T.F.* (in Russian) **14**, 270.
Starobinsky, A. A. and Churilov, S. M.: 1973, *Zh.E.T.F.* (in Russian) **65**, 3.

SCALAR WAVES IN THE EXTERIOR OF A SCHWARZSCHILD BLACK HOLE

S. PERSIDES

Astronomy Dept., University of Thessaloniki, Thessaloniki, Greece

Abstract. Fourier and Laplace transforms are used to study rigorously the properties of a test scalar field Ψ in the exterior of a Schwarzschild black hole of the mass m. In the Fourier analysis we examine the properties of the solutions of the radial wave equation and the relations of the exterior and interior solutions of the following four cases: (i) $\omega \neq 0$, $m \neq 0$, (ii) $\omega = 0$, $m \neq 0$, (iii) $\omega \neq 0$, $m = 0$, (iv) $\omega = 0$, $m = 0$.

In the Laplace analysis we show rigorously the following theorem: *If* $\Psi(t, r, \theta, \varphi)$ is the field of a point test particle falling into the black hole,

$$[\partial \Psi / \partial t]_{t<t_0} = 0,$$

and $\lim \Psi$ exists, *then* $\lim \Psi = 0$. The proof of this theorem is based on the facts that (a) $t + 2m \ln(r - 2m)$ is finite for the particle even on the horizon, and (b) the behavior of Ψ as $t \to +\infty$ is related to its Laplace transform near the origin of the complex plane.

References

Persides, S.: 1973, *J. Math. Phys.* **14**, 1017.
Persides, S.: 1974, *J. Math. Phys.*, to appear.

ELECTROMAGNETIC WAVES IN THE EXTERIOR OF A SCHWARZSCHILD BLACK HOLE

H. STEPHANI and E. HERLT

Sektion Physik, Universität Jena, G.D.R.

Abstract. Using the technique of Debye potentials, the regular (nonstatic) EM fields are found to behave at $r = M$ either as purely ingoing or as purely outgoing waves. After separation the radial equation is treated in the short wavelength approximation.

THE ROLE OF GRAVITATIONAL RADIATION IN THE EVOLUTION OF DWARF NOVAE

JOHN FAULKNER
Lick Observatory, U.S.A.

Abstract*. Mechanisms promoting mass transfer include (i) envelope instability or (ii) nuclear evolution of the red star and (iii) gravitational radiation of orbital angular momentum. Growing observational evidence against (i) is supported by recent theoretical work on the medium and long-term response of stellar radii to mass-loss (Eggleton, Faulkner and Webbink, in progress). Since (ii) is in most cases too slow a process, (iii) is left as the best surviving explanation.

* See proceedings of IAU Symposium No. 66.

ON THE DESCRIPTION OF HIGH-FREQUENCY GRAVITATIONAL WAVES

M. A. H. MacCALLUM

King's College, Cambridge, Great Britain

Abstract. Taub and I have used the 'Average Lagrangian' method (Whitham, 1971; Dougherty, 1970) for the case of gravitational waves (MacCallum and Taub, 1973; Taub, 1973). A difficulty arises in going beyond first-order effects as gauge waves and physical waves appear to interact (MacCallum, 1973). The source of this difficulty is not yet known, but doubt is cast on the usual conceptual framework.

References

Dougherty, J. P.: 1970, *J. Plasma Phys.* **4**, 761.
MacCallum, M. A. H.: 1973, paper in preparation.
MacCallum, M. A. H. and Taub, A. H.: 1973, *Comm. Math. Phys.* **30**, 153.
Taub, A. H.: 1973, *Proceedings of C.N.R.S. Colloque*, Paris.
Whitham, G. B.: 1971, in A. H. Taub (ed.), *Studies in Applied Mathematics*, vol. 7, Mathematical Association of America Studies in Mathematics, Prentice Hall, Englewood Cliffs, N.J.

ALTERNATIVE APPROACH TO INFINITY

PETER G. BERGMANN

Syracuse University, U.S.A.

Abstract. Following Penrose's construction of space-time infinity by means of a conformal construction, in which null-infinity is a three-dimensional domain, whereas time- and space-infinities are points, Geroch has recently endowed space-infinity with a somewhat richer structure. An approach that might work with a large class of pseudo-Riemannian manifolds is to induce a topology on the set of all geodesics (whether complete or incomplete) by subjecting their Cauchy data to (small) displacements in space-time and Lorentz rotations, and to group the geodesics all of whose neighborhoods intersect into equivalence classes. The quotient space of geodesics over equivalence classes is to represent infinity. In the case of Minkowski, null-infinity has the usual structure, but I^0, I^+, and I^- each become three-dimensional as well.

THE GEODETIC INTERVAL IN A RIEMANNIAN SPACE-TIME IN THE SECOND POST-MINKOWSKIAN APPROXIMATION*

REINER WILHELM JOHN

Zentralinstitut für Astrophysik der Akademie der Wissenschaften der D.D.R., Potsdam-Babelsberg, G.D.R.

Abstract. The knowledge of the geodetic interval between two points in a Riemannian space-time with the metric g_{ab} is essential for statements on the time delay in a gravitational field represented by g_{ab} and makes possible to derive explicit criteria for clear-cut wave propagation. The nonlinear differential equation for the geodetic interval is integrated via perturbation expansion in the second post-Minkowskian approximation.

* To be published.

MAGNETIZATION, MATTER-ANTIMATTER SYMMETRY AND THE BARYON-PHOTON RATIO IN THE UNIVERSE

M. A. MELVIN

Physics Dept., Temple University, Philadelphia, Pa. 19122, U.S.A.

Abstract. It is shown that the universal magnetic field, or sufficiently extended intergalactic fields, of magnitude $\gtrsim 10^{-9}$ G would have aligned the magnetic moments of all leptons at an early time. Unless an upper limit to temperature exists, the alignment of all nucleons would also have occurred at an earlier time when the temperature was $m_N/m_e \sim 2000$ times higher. Possible inferences of this early magnetization of the matter in the universe for observation are discussed. The one selected for particular analysis is the parameter

$$\frac{\text{No. baryons}}{\text{No. photons}} = \eta \approx 10^{-9}$$

expressing the inverse 'hotness' of the universe. The matter-anti-matter symmetric theory of η given by Omnes, amended by Steigman and Kundt is reviewed. The effect of the large scale magnetic alignment on the value of η resulting from the annihilation era is then discussed by means of a model in which matter and antimatter droplets or filaments are in quasi-equilibrium under magnetization and effective surface forces. The magnetized droplets affect the diffusion of neutrons, which is dominant in the annihilation era.

A NEW GENERAL COVARIANT APPROACH TO THE GENERAL RELATIVISTIC TWO-BODY PROBLEM*

ARNOLD ROSENBLUM

Astronomische Institute der Universität Bonn, F.R.G.

Abstract. A new general covariant approach to the general relativistic equations of motion is presented. It is stressed that our present understanding of the development of binaries due to general relativistic effects and of the power emitted by these systems in the form of gravitational radiation is highly unsatisfactory.

* To appear in *Phys. Rev.*

GRAVITATIONAL DEVIATION REACTION

TERRENCE J. SEJNOWSKI

University of California at Santa Barbara, Calif., U.S.A.

(Presented by J. B. Hartle)

Abstract. A gravitational radiation reaction tensor is calculated for a point quadrupole source in the fast motion approximation. The calculation is in direct analogy with Dirac's original calculation of electromagnetic radiation reaction. In the gravitational case a tidal deviation reaction tensor results which is linear in the fifth derivatives of the source quadrupole moments.

TIDAL TENSOR AND THE EMISSION AND ABSORPTION OF GRAVITATIONAL RADIATION

TERRENCE J. SEJNOWSKI

University of California at Santa Barbara, Calif., U.S.A.

(Presented by J. B. Hartle)

Abstract. The absorption and emission of gravitational radiation can be calculated in the long wavelength limit by use of the *tidal tensor*, defined as the gradient of the gravitational pseudoforce,

$$\begin{pmatrix} \text{local 'force'} \\ \text{on } T_{ab} \end{pmatrix} \equiv (\sqrt{-g}\, T^{ab})_{,b} = f^a ;$$

thus

$$(\text{tidal sensor})^a{}_d \eta^d \equiv f^a{}_{,d}\eta^d = \tfrac{2}{3} R^a{}_{(bc)d} T^{bc}\eta^d + O(\eta^2),$$

where η^b is orthogonal to the 4-velocity u^a, and normal coordinates are understood.

The exchange of energy and momentum between an extended body and the gravitational field is governed by appropriate integrals of the tidal tensor over space and time.

The tidal tensor is the trace $T^a{}_{bd}{}^b$ of the Bel-Robinson tensor T_{abcd}. In emty space this trace is zero and the tidal tensor vanishes; there is no local exchange of energy.

COMPLEX MAXWELL AND EINSTEIN FIELDS

EZRA T. NEWMAN*
University of Pittsburgh, U.S.A.

Abstract. We consider the class of regular (in a certain precise sense) null vector fields, l_μ which have the following properties; they are (1) tangent to geodesics, (2) diverging, (3) shear free, (4) twist (or curl) free. It is well known that the vacuum Einstein fields whose principle null vector field (pnvf) satisfies (1)–(4) are the Robinson-Trautman (1962) (RT) metrics and those which satisfy (1)–(3) are the algebraically special twisting metrics, (Kerr, 1963). To understand these metrics better we ask for those Maxwell fields (in flat space) whose pnvf also satisfy conditions (1)–(4) and (1)–(3). It can be shown that (1)–(4) imply (and are implied by) that the Maxwell field is a Lienard-Wiechart (LW) field. (This establishes the analogy between the RT metrics and the LW fields.) Conditions (1)–(3) imply that the Maxwell field is a complex LW field. (We mean by this that if the Maxwell equations are complexified (Newman, 1973) (in complex Minkowski space) then the real solution in question is induced from the complex solution which is associated with a charged particle moving along an arbitrary *complex* world line.) Finally it can be shown that the Einstein equations can be complexified and that the algebraically special twisting metrics can be interpreted as if they had a point source moving in the *complex* manifold and are thus analogous to the complex LW fields.

References

Kerr, R. P.: 1963, *Phys. Rev. Letters* **11**, 237.
Newman, E. T.: 1973, *J. Math. Phys.* **14**, 102.
Robinson, I. and Trautman, A.: 1962, *Proc. Roy. Soc.* **A265**, 462.

* This work was done in collaboration with R. Lind.

ON BLACK AND WHITE HOLES

M. A. MARKOV

Laboratory of Theoretical Physics, Joint Institute for Nuclear Research, Dubna, U.S.S.R.

Abstract. Various possible cases of spherically systems the matter of which is localized in a domain smaller than the corresponding gravitational radius is considered.

The metic of the Friedmann closed world or of a part of it with an external continuation is suggested as a model of these systems.

There can exist black holes which are described by semi-closed metrics (black holes of the second kind). The class of systems in question may be both in the state of collapse and in the state of anti-collapse (including the state of 'white holes').

There are some grounds to suppose that collapse of celestial bodies should stop in the domain $\hbar/m_v c$, where m_v is the mass of the vector meson, and that the pair production effect due to collapse of a charged sphere should conserve the Laplace determinism of the process.

The role of the charges of sources of different fields (electromagnetic, meson vector, scalar long-range, scalar meson, various versions of neutrino fields) in the deformation of the external and internal metric of black and white holes is analysed.

In this consideration a number of problems arises (the absence of horizon in the case of any small charges of scalar fields, the presence of the generalized Gauss theorem for vector meson field etc.), which provide evidence that the assertion 'Black hole has no hair' needs further investigations. In particular, the inverse process of formation of hair (e.g. vector-meson, scalar fields) in the process of anti-collopose has not been studied yet.

For the limiting case of the Nordström-Reissner metric $m=e$ (more correctly $m \to e$) two essentially different possibilities of continuing to the internal metric are considered (the Papapetrou case and the case which we called 'friedmon metric' describing charged black holes of the second kind (friedmons)).

In the case of charged holes of the second kind (friedmons) the occurance of quantum effects (pair productions) can reduce the horizon surface and violate the Hawking theorem.

The notion of black holes may turn out to be essential in elementary particle theory: among the intermediate states in elementary particle theory there are states the characteristic feature of which is the localization of arbitrary large energies (masses) in a domain smaller than the gravitational radius.

Collapse and anticollapse of material systems have long been the object of theoretical investigations. At the present time when astrophysics becomes gradually an experimental science the interest in such systems increases greatly. Much attention is also paid to peculiar changes in the process of collapse of the global properties of matter which are being extensively discussed. In particular, a possible disappearance of some fields in the external metric of similar systems in the process of collapse (when the collapsing matter is behind the Schwarzschild sphere) is widely discussed.

This situation figuratively is defined by Wheeler (1971) as follows: 'A black hole has no hair'.

In what follows we are dealing with the systems in which matter is localized in regions smaller than the corresponding gravitational radii. The state of these systems has a certain variety. It can, in particular, be a both a collapse and an anticollapse. The systems under discussion have also other differences if sources of different fields are included into them. It seems advisable to make a certain classification of possible objects of this kind in the framework of general relativity. In what follows we restrict ourselves in the main to the consideration of the systems when the appropriate metric in the region of localization of gravitating matter is described by a linear element of the closed Friedmann world (Hoyle et al., 1965), and the external solution matched with this 'internal' solution at a given moment is the Schwarzschild solution. We exclude thereby a very interesting class of external metrics of the Kerr type because the appropriate internal metrics matched with them have not been constructed yet. In the discussion of the fate of short-range forces in the external metric of these systems it is advisable to employ also systems of small total masses. Such systems are, in principle, allowed by theory. The gravitational radius of the systems in question may be e.g. of the order $\hbar/m_v c$, where m_v is the mass of, e.g., a vector meson. The quantity $\hbar/m_v c$ is of the order 10^{-13} cm: here we are still in the range of applicability of the non-quantum theory of gravitational field. In other words, the given length $\hbar/m_v c$ is by 20 orders of magnitude larger than the corresponding gravitational length $(l_{gr}=(\hbar c \varkappa)^{1/2} c^{-2} \sim 10^{-32}$ cm), where we may suspect the invalidity of the classical Einstein gravitational theory. In addition, we may always, if needed, put formally m_v to be very small, i.e. $\hbar/m_v c$ to be arbitrary large. There is no real physical sense of considering short-range external fields of celestial bodies at distances $\hbar/m_v c \sim 10^{-13}$ cm and discussing whether it is possible to detect them experimentally. It should be stressed that systems with gravitational radius of an order of 10^{-13} cm are by no means related to microworld objects. Such a gravitational radius is assigned to the mass

$$M_0 \sim \frac{\hbar}{m_v c} \frac{c^2}{2\varkappa} \sim 10^{14} \text{ g} = 10^8 \text{ ton}. \tag{1}$$

The Friedmann linear element

$$ds^2 = a^2(\eta) \, d\eta^2 - a^2(\eta) \, d\chi^2 - a^2(\eta) \sin^2 \chi (d\theta^2 + \sin^2 \theta \, d\varphi^2)$$

or

$$ds^2 = c^2 \, dt^2 - a^2(t) \{d\chi^2 + \sin^2 \chi (d\theta^2 + \sin^2 \theta \, d\varphi^2)\} \tag{2}$$

is known to describe one of the models of the closed world.

Here the variable χ changes in the limits

$$0 \leqslant \chi \leqslant \pi, \qquad a(\eta) = a_0(1 - \cos \eta). \tag{3}$$

The variable η is connected with the time t by a simple relation:

$$t = \frac{a}{c}(\eta - \sin \eta).$$

The same Equation (2) can describe the internal metric of the so-called black hole, more strictly, of a certain model of the black hole, provided the matter of the system is distributed in such a manner that the region χ is filled only to $\chi_m \leqslant \pi/2$. And then the external metric, e.g. (Euclidean at infinity), should be 'matched' with the internal Friedmann metric in an appropriate manner. The black hole is usually viewed as the final state of a collapsing system when the matter of the system is behing the

Schwarzschild sphere. We imply here the state which the collapsing system tries to reach at the Schwarzschild time increasing to infinity. We call this limiting state of the system black hole of the first kind.

We may define strictly this object in the framework of the abovementioned metric with the above-mentioned restriction on the range of the χ values. Namely,

(1) Black hole of the first kind:

$$0 < \chi_m \leq \frac{\pi}{2}. \qquad (4)$$

As far as the metric is nonstatic and the question is to discuss collapse, further fate of the system is related to decreasing $a(t)\,\chi_m$ value, i.e. to reducing sizes of the system which 'is' already behind the Schwarzschild sphere.

The variety of the objects described by the internal metric (2) is not limited by the state-black hole of this kind.

In fact, if matter fills in the region χ so that $\pi/2 < \chi_m < \pi$ then there arises a space with a semi-closed metric. This space has the following properties. If spheres or $\chi > 0$ are circumscribed around the point $\chi = 0$ at a given moment then the surface of the sphere is

$$S = 4\pi a^2(t) \sin^2 \chi. \qquad (5)$$

The surface of the sphere S increases with increasing χ to $\chi = \pi/2$. However, when $\chi > \pi/2$, the sizes of the sphere decrease and for $\chi = \pi$ the sphere reduces to a point which implies that the world becomes closed.

Semi-closed worlds were first studied by Klein (1961), Zel'dovich (1962), and Novikov (1962). Klein called the system of this kind Friedmann world with external continuation '(Mit Außenwelt)'.

When $\chi < \pi/2$ (black hole of the first kind) the surface of the spheres in question increases monotonically with increasing χ at a given moment t. In the external metric the quantity $a(t) \sin \chi = r$ assumes also the meaning of a monotonically increasing radius. The semi-closed metric (when $\pi/2 < \chi_m < \pi$) is characterized by the existence of a minimal surface, a minimal value of r in vacuum, where $\partial r/\partial \chi = 0$ and $\partial^2 r/\partial \chi^2 > 0$ (for details see Markov and Frolov, 1970; Markov, 1971).

In other words, the semi-closed metric is characterized by the presence of a specific throat which links the internal and external metric.

In principle, among celestial bodies there may be objects of the second class just mentioned. Thus

(2) Objects with semi-closed metric: $\pi/2 < \chi_m < \pi$ are black holes of the second kind.

Although the objects of the second kind are, in their astrophysical nature, analogous to the black holes, they differ essentially from the black holes of the first kind. The objects of the former class can arise (for instance) in the process of evolution of white holes (see Appendix II).

Finally, the third case:

(3) $\chi_m = \pi$ – the Friedmann closed world.

The matter density $\mu(t)$ integrated over the whole space of a closed world gives a 'bare' mass of the system i.e. the total mass without taking into account the gravitational defect:

$$M_0 = 2\pi^2 \mu(t)\, a^3(t). \tag{6}$$

This value of the 'bare' mass defines the sized of the radius of the closed world at the moment of its maximum extension:

$$a_0 = \frac{\varkappa M_0}{3\pi c^2}. \tag{7}$$

The latter expression is immediately obtained from the Einstein equation

$$\left(\frac{\dot{a}}{a}\right)^2 = \frac{2\pi}{3}\varkappa\mu - \frac{c^2}{a^2}. \tag{8}$$

If we put in Equation (8) $\dot{a} = da/dt = 0$, then according to Equation (6), $\mu_0 = M_0/2\pi^2 a_0^3$.

These relations show that a *closed world can, in principle, contain an arbitrary small amount of matter and has arbitrary small sizes*. It appears that quantum mechanichs alone can impose here definite restrictions. Thus, for

$$M_0 = \left(\frac{\hbar c}{\varkappa}\right)^{1/2} \sim 10^{-5}\, \text{g}$$

the dimensions of the world

$$a_0 \sim \frac{(\hbar c \varkappa)^{1/2}}{c^2} \sim 10^{-33}\, \text{cm}. \tag{9}$$

There are some grounds to believe that at these distances the notion of length losses its meaning due to quantum fluctuations of the metric (Regge, 1958; Blokhintsev, 1960).

The total mass of a part of a closed world, localized in the region from $\chi = 0$ to χ_m (i.e. the bare mass minus its gravitational defect) is given by the expression

$$M_{\text{tot}} = \frac{c}{\varkappa} a_0 \sin^3 \chi_m. \tag{10}$$

(Klein, 1961; Zel'dovich and Novikov, 1967).

Thus, the total mass of the closed system ($\chi_m = \pi$) is zero.

The semi-closed system is characterized by two lengths. One of the lengths l_{in} characterizes the internal sizes of the system

$$l_{in} = a(t)\,\chi. \tag{11}$$

The maximum value of the internal sizes may be arbitrary large

$$l_{in}^{\max} = a_0 \chi_m, \tag{12}$$

where a_0 is the radius of the system defined by the bare mass, according to Equation (7).

The other sizes are associated with the surface which surrounds a part of the Friedmann world filled in with matter. This is the surface of the 'throat' $S = a^2 \sin^2 \chi_0$ with its radius

$$l_{out} = a(t) \sin \chi_0, \qquad \chi_0 > \chi_m, \qquad (13)$$

which, depending on the χ_0 value, may be arbitrary small. At $\chi_0 \to 0$, l_{out} tends to zero. The quantity l_{out} is the external dimension of the system, that is the dimensions of the system which an external observer sees.

Systems like our Universe with numerous galaxies having ultramacroscopic internal dimensions (l_{in}) may have arbitrary small external dimensions (l_{out}). Even this surface may be microscopic dimensions and, according to Equation (10), may possess a microscopic total mass*. In this sense, we may speak of a peculiar relativity of the notions of macro and micro (Markov, 1971).

Further variety of the objects under discussion may be associated with the fact that they can be in two different states defined by the initial conditions. In other words, the internal dimensions of the system can eventually either decrease

$$\left(\frac{da}{dt} < 0 - \text{collapse} \right)$$

or increase

$$\left(\frac{da}{dt} > 0 - \text{anticollapse} \right).$$

These objects being in the state of collapse are systems absorbing matter from the surrounding space (black holes), while in the state of anticollapse there may, in principle, be radiating objects sources emitting matter (white holes).

Strictly speaking, so far we are not aware of whether black holes exist and whether the collapsing systems are realized in nature.

However, we may apparently assert with a great probability that there exists, at least, one system in the state of anticollapse, this is precisely our Universe. It is quite probable that our Universe is just a white hole. Very little is known about the end fate of the collapsing system.

It is hard to say *a priori* whether it is possible to exclude the existence of periodic states when a collapse is changed by an anticollapse. In principle, such a possibility may be given by, e.g., long-range fields of the type of electrostatic fields (Novikov, 1966). Short-range vector meson fields may play the same role provided in the process of collapse the system turns out to be in the region $\hbar/m_v c$ (Berezin and Markov, 1969), where m_v is the vector meson mass.

But studying this situation we see that the collapse changes by an anticollapse

---

* In the framework of the classical theory there are not restrictions on the smallness of the total mass. It may be, e.g. of the order of the mass of an elementary particle.

which however occurs not in the space where the collapse took place (Novikov, 1966; de la Cruz and Israel, 1967; Bardeen, 1968).

On the one hand, this unexpected and unusual situation forces us to search for the explanation may be in an insufficiently real description of the process (for instance, in neglecting e.g. the huge pair production effect in big meson fields of the final stage of collapse (see Appendix 1). On the other hand, the same situation gives a possibility of making nontrivial speculations concerning topological structure of our Universe (Sakharov, 1971). Should we exclude in this case the possibility that some of the objects in question might enter in this state of anticollapse in the state of a white hole only recently, as a result of its periodic development, and this might happen precisely in our space.

These considerations would suggest an interpretation of the formation of galaxies in the spirit of the conceptions of Ambartsumian (1962) which, in his opinion, follow from astronomical observations. A similar hypothesis was proposed by Novikov (1964) and Ne'eman (1965). But finally, the problem reduces to different formulations of the appropriate initial conditions.

It is obvious that spontaneous emergence of a white hole in our space is not a very easy understandable event. However, in a certain sense the same problem arises when discussing the original moment in the development of our Universe.

Objects with semi-closed matric can be classified as an independent class of objects.

A black hole of the first kind cannot turn to, e.g. a semi-closed system. The matter is that the black hole can only absorb matter from the surrounding space, can only increase its gravitational radius, while the formation of a system with semi-closed metric requires for the Schwarzschild sphere to decrease its sizes.

Another situation arises if we deal with objects in the state of anticollapse (see Appendix 2).

Up to this point of our presentation the subject of our consideration were electrically neutral systems. The introduction of even arbitrary small electric charges in the systems in question changes essentially the situation and thereby changes the classification of the objects we are interested in.

First of all, the third case turns out to be impossible – system with closed metric. The closed Friedmann metric perturbed by the presence of an arbitrary small electric charge is the case of the semi-closed metric.

In the Friedmann metric (2) we can formally find a solution for the electrostatic potential φ, for simplicity at the moment of maximum extension of the world.

Such a solution is of the form

$$\varphi = \frac{\text{const}}{a \sin \chi}. \tag{14}$$

As $\chi \to \pi$ the expression for the potential becomes infinite. For $\chi = \pi$ there appears a particular feature characteristic of a point source the presence of which at this place has not been supposed by us. This is precisely the form in which the contradiction between the closed metric and the attempt to introduce in this metric the total non-zero electric charge is revealed. In fact, at the place where $\chi \to \pi$ there arises something representing the image of the field source localized at $\chi = 0$. The contradiction with the closed metric disappears if we imply the charge of an opposite sign. Then the total electric charge of the system turns out to be equal to zero. If the strength lines go out from the charge at the point $\chi = 0$, then they must end in this case on the charge of an opposite sign at the point $\chi = \pi$.

The character of the deformation of the closed world metric by a small electric charge was considered in detail by Markov and Frolov (1970). When the electric charge is arbitrary small a noticeable deviation from the Friedmann metric arises only for χ arbitrary close to π.

In other words, if we circumscribe spheres with $\chi > 0$ around a small charge ε localized at $\chi = 0$ then these spheres are characterized by the expression (5) up to χ very close to π.

In the domain $\chi > \pi/2$ the spheres will greatly decrease with increasing χ, as in the absence of the electric change. However, in this case (at $\chi > \pi/2$) the spheres play the role of peculiar focusing lenses for strength lines of an electrostatic field.

A detailed study shows that when the density of strength lines ('hair' of the electromagnetic field) becomes such that the corresponding value for the electrostatic potential reaches a value close to

$$\varphi = \frac{c^2}{\varkappa^{1/2}}, \tag{15}$$

then with further increase of χ the spheres in question begin growing again and the metric turns to a particular case of the well-known Nordström-Reissner metric. This particular case is characterized by the Schwarzschild mass

$$M_{tot} = \frac{\varepsilon}{\varkappa^{1/2}} \tag{16}$$

and the metric itself (outside throat) is

$$ds^2 = \Phi^2 c^2 \, dt^2 - \frac{dr^2}{\Phi^2} - r^2(d\theta^2 + \sin^2\theta \, d\varphi^2), \tag{17}$$

where

$$\Phi = \left(1 - \frac{\varepsilon \varkappa^{1/2}}{c^2 r}\right)^2. \tag{18}$$

The radius of a minimal sphere which is allowed by the electrostatic charge is proportional to the electric charge

$$r_{min} = \frac{\varepsilon \varkappa^{1/2}}{c^2}. \tag{19}$$

The contradiction with the closed metric is 'created' by the Gauss theorem. Figuratively speaking, the electric strength lines ('hair') become more dense for χ close to π

so that they 'punch' in the closed metric a 'wormhole' (throat) into which the electric vector flux rushes forming outside the given material system a particular case of the Nordström-Reissner metric. Outside the throat this metric is quite analogous to the metric suggested by Papapetrou (1945).

But the physical object characterized by this metric and the complete space-time description of the metric at small distances ($r < \varepsilon \varkappa^{1/2}/c^2$) are of quite different nature. Special attention should be paid to this fact.

The Papapetrou metric describes the static case when gravitational attraction of particles is equilibrated by electrostatic forces of repulsion of their electric charges. Roughly speaking, from the equality $\varkappa m^2/r = e^2/r$ it follows the relation (16) $m = e/\varkappa^{1/2}$.

A detailed analysis (Markov and Frolov, 1972) of the Papapetrou metric shows that in this case as one should except for the static case, matter cannot be behind the Schwarzschild sphere. In the Papapetrou model the sizes of the domain in which matter is localized is necessarily large than the gravitational radius of the system.

In our case we are dealing with the system which is inside the Schwarzschild sphere. In this model the metric is nonstatic. The external metric in the form (18) is a limiting case of the Nordström-Reissner metric for $M > \varepsilon/\varkappa^{1/2}$ when $M \to \varepsilon/\varkappa^{1/2}$.

The general case of the Nordström-Reissner metric is known to be described by the expression (17) as well, but now Φ is of the form

$$\Phi^2 = 1 - \frac{2\varkappa m_0}{c^2 r} - \frac{\varkappa \varepsilon^2}{c^4 r^2}. \tag{20}$$

The expression (20) has two roots

$$r_\pm = \frac{\varkappa m_0}{c^2} \left\{ 1 \pm \left(1 - \frac{\varepsilon^2}{\varkappa m_0^2} \right)^{1/2} \right\}. \tag{21}$$

In our limiting version of this metric, according to (16) $M_{\text{tot}} = \varepsilon/\varkappa^{1/2}$, r_+ and r_- are equal to each other.

The physical picture of the Friedmann metric distorted by the presence of the small charge differs from the corresponding picture of the Papapetrou model by that in the latter the values of the total mass of the system and its bare mass coincide. While in our case the total mass is of essentially electrostatic nature. The value of the bare mass (say, the number of nucleons in the system) may be arbitrary. Here the bare mass is completely cancelled by the gravitational defect of the system. The internal metric together with the external metric (18) which at $\varepsilon \to 0$ transforms to the metric of the Friedmann closed world is given by us the name of friedmon metric and the object itself – 'friedmon'.

We have considered in detail the friedmon metric and have introduced for the given object a special term 'friedmon' because in literature the case $\varepsilon^2/\varkappa \to M^2$ is interpreted as the Papapetrou case, or which is the same, as the Bonnor case (1960) even when one considers an entire analytic continuation of the metric for $\varepsilon^2/\varkappa = M^2$ (Carter, 1966). According to (19) the radius of the throat increases proportionally to the total electric charge.

This is the description of the throat from the point of view of classical (nonquantum) physics. But from the viewpoint of quantum physics the state of the throat cannot be stable.

Indeed, if at some initial moment a throat with the above mentioned properties arises then in its superstrong electrical field* there inevitably occurs a violent process of production of any kind of electrically charged pairs: proton-antiproton pairs, any kind of meson pairs, and finally, electron-positron pairs. Charges of opposite signs will attempt to decrease the effective charge of the throat while the charges of the other component of pairs will flow to the euclidean infinity.

The decrease of the electric charge of a system due to the production of charged particle pairs results in the decrease of the electric charge of the throat. The peculiarity of the quantum effect of electron-positron pair production consists in that the pair components may be spaced by a considerable space-like interval (Zel'dovich, 1971). *It may happen that the electron and positron are created on opposite sides from the event horizon.* A 'subbarrier' penetration of particles with opposite charges through the event horizon (throat), particle pairs produced outside the throat, is also possible.

From the viewpoint of the Schwarzschild observer a particle approaches the gravitational radius of the black hole during an infinitely long time. But in this process there is a striking peculiarity which should be borne in mind in a number of facts.

The law of change of this spacing with time is given by the formula (Zel'dovich and Novikov, 1967).

$$r = r_{gr} + (r_1 - r_{gr}) e^{-c(t-t')/2r_{gr}},$$

here r is the location of the particle at a moment t', r_{gr} is the gravitational radius, $(r_1 - r_{gr}) \ll r_{gr}$.

Here the numbers are very instructive. In fact, let $r_{gr} \sim 10^5$ cm, and $(r_1 - r_{gr}) \sim 10^3$ cm.

As is seen from the formula, even this short way, amount 10 m, is covered by the particle during an infinite time. The main thing is however that in a second the distance between the Schwarzschild sphere and the test particle becomes infinitely small:

$$\Delta r \sim 10^3 \, e^{-10^5} \text{ cm} < 10^{-10000} \text{ cm},$$

which is infinitely small even compared with the characteristic length of quantum fluctuations of the metric $l \sim 10^{-33}$ cm.

Because of the presence of huge fields in the systems in question (friedmons) near the throat the pair production process under given initial conditions occurs mainly directly in the throat region. *But spontaneous appearance of a charge of an opposite sign 'inside' the fridmon decreases the internal charge of the friedmon.* Consequently, according to (19), the radius of the throat and the horizon surface, the Schwarzschild sphere surface, decrease as well.

In other words, in quantum domain quantum effects can violate the Hawking

* The value of the potential $\varphi = c^2/\sqrt{x}$ appears to be the maximum one which is allows by nature in R space. This expression contain no electric charge (!!).

theorem. Another case of quantum violation of the Hawking theorem was recently indicated by Bekenstein (1973). If we suppose that the electron of a pair produced near the throat carried away at infinity an energy $E = \varepsilon e/r_{gr} = e\varkappa^{-1/2}c^2$ when it pushes off from the throat charge (ε), then the remainder conserves the characteristic property of the friedmon:

$$M' = \frac{\varepsilon'}{\varkappa^{1/2}}.$$

Thus, in the process of spontaneous pair production near a given friedmon its total charge, total mass and the sizes of the throat can decrease. This process has not been studied yet in detail, but the estimates (Markov and Frolov, 1970) show that the pair formation process results in a decrease of the friedmon charge in the final state to a value $Ze < 137\,e$. According to Landau (1955), further vacuum polarization appears to be capable of decreasing the charge to $\varepsilon = 1$. It is essential that the same value of the electric charge of the final state arises for any charge value of the initial friedmon. In any case, the latter assertion holds for relatively small initial Z, namely for $Z < 10^{20}$ for initial

$$r_h = \frac{\varepsilon \varkappa^{1/2}}{c^2} \sim 10^{-14}\text{ cm}.$$

The appearance of a friedmon state in the process of evolution of white holes accompanied by radiation of charges was considered by Frolov (1973) (see Appendix 2).

If in this process the Hawking theorem is not violated then the mass of the initial system may be only half of the original mass. It is however quite possible that the whole process occurs under these conditions with a complete violation of the Hawking theorem.

In any case a friedmon with electric charge equal to unity is a stable object of friedmon metric. The parameters of this friedmon are as follows: mass $m_f^e \sim 10^{-6}$ g, sizes $r_f^e \sim 10^{-33}$ cm.

If there exist baryon hair of black holes then in addition to the electrostatic friedmon there might be meson friedmons:

$$m_f^g = \frac{g}{\varkappa^{1/2}} \sim 10^{-5}\text{ g}, \qquad r_f^g = \frac{g\varkappa^{1/2}}{c^2} \sim 10^{-32}\text{ cm},$$

where $g^2/\hbar c \sim 1$, g is the baryon charge, the source of a vector meson field (Markov, 1970). It should be stressed once more that r_f^e and r_f^g are the external sizes of the system. Its internal sizes $a(t)\chi_0$ and its bare mass, the number of nucleons, may be expressed, e.g. by ultramacroscopic numbers. The particles under discussion and the particle characterized by the constants $\hbar$, c, $\varkappa$, i.e. $(\hbar c/\varkappa)^{1/2} \sim 10^{-5}$ g – Plank-Wheeler particle – may be regarded as possible 'elementary' particles of limiting maximum large masses, that is they can form a group of 'maximon' (Markov, 1965). Maximons may be of a relict origin, i.e. may exist in our Universe perpetually. Hawking (1971) has also assumed the existence of such particles with charge ~ 30. But the object of such small sizes ($\sim 10^{-32}$ cm) with such a charge will turn out to be unstable due to vacuum polarization.

Thus, we see that the presence of electric charges, sources of the electrostatic field changes essentially the classification of objects the internal metric of which is partially or completely characterized by the metric of Friedmann closed world. In the latter case there naturally and inevitably arises a semi-closed metric. There arises the question of what the effect of the change of the closed world metric is if it is perturbed by small charges of other fields: for example, scalar, massive vector fields or fields related to weak interactions.

Numerous papers assert that the sources of other fields (for the exception of electromagnetic field) localized in matter inside the Schwarzschild sphere (hole) do not excite appropriate fields outside given objects: 'Black hole has no hair'. But here many problems arise which are still open.

1. The Meson Vector Field

At first glance it seems that, as in the case of electrostatic, the presence of sources of the vector meson field is incompatible with the closed metric.

Indeed, using the standard procedure, it is easy to obtain, at the moment of maximum expansion, for the meson vector field potential the following expression

$$\varphi_0 \sim \frac{\beta e^{-\lambda \chi}}{a_0 \sin \chi}, \qquad \lambda = (a_0^2 m_v^2 - 1)^{1/2}, \qquad (22)$$

where m_v is the vector meson mass. This expression can naturally be thought of as an analog, in the euclidean space, to the ordinary expression $\varphi_0 \sim \beta(e^{-m_v r}/r)$. On the basis of (22), it might be concluded that the potential φ_0 at $\chi \to \pi$ is divergent as in the electrostatic case that the presence of vector meson field sources with non-zero total charge is incompatible with the closed metric, and that the given system must have a continuation – an external metric, i.e. an external meson vector field.

But this assertion is wrong: it is based on the use of the boundary conditions which are ordinary for the euclidean space. These conditions of finiteness of the solution select solutions exponentially damping with increasing r.

In the case of a closed world there is no spacial infinity, and therefore a more general type of the solution is possible (Markov and Frolov, 1973):

$$\varphi_0 = \frac{\beta e^{-\lambda \chi}}{a \sin \chi} + \frac{\gamma e^{+\lambda \chi}}{a \sin \chi}. \qquad (23)$$

The requirement that the solution should be finite and continuous is satisfied by the condition imposed on the coefficients β and γ:

$$\beta e^{\lambda \pi} + \gamma e^{-\lambda \pi} = 0. \qquad (24)$$

Thus, outside the point source g_0 the solution is

$$\varphi = \frac{g_0 \operatorname{sh} \lambda(\pi - \chi)}{a \operatorname{sh} \lambda \pi \sin \chi}. \tag{25}$$

Now φ_0 at $\chi \to \pi$ is finite, and the field $\partial \varphi_0 / \partial \chi$ at $\chi \to \pi$ vanishes.

Thus, we are led to the conclusion that the closed world may contain the total nonzero baryon charge (vector meson field source). This establishes the essential difference of the massless vector field (i.e. electrodynamics) from the meson vector field. The obtained results do not contradict the assertion that the meson vector field is absent outside black holes. At the same time, it is not a proof in favour of this assertion.

Meanwhile, as in the case of electrodynamics, the previous consideration is a rigorous proof of the presence of an electrostatic field sources are found behind the Schwarzschild sphere the vector meson field outside the Schwarzschild sphere vanishes (Bekenstein, 1972; Teitelboim, 1972; Thorne, 1971).

It is worth noting that, in the case of the meson vector field as well, there is a certain peculiar analog of the Gauss theorem, more correctly, its peculiar generalization. More detailed consideration of the available relations leads us to the conclusion that the situation is really more complicated than seems at first.

We consider a mesodynamic analog of the Gauss electrodynamic theorem in an Euclidean metric, for simplicity (Markov, 1970).

Let a baryon charge ϱ be localized in a certain domain so that

$$\varrho \neq 0, \quad r < r_0,$$
$$\varrho = 0, \quad r > r_0.$$

For the mesodynamic vector E_n flux through a closed surface surrounding the charge, on the basis of the equation

$$F^{\mu\nu}{}_{,\nu} - m_v^2 \varphi^\mu = -4\pi j^\mu, \quad j^4 = \varrho, \tag{26}$$

we get the following expression

$$\int E_n \, ds = m_v^2 \int \varphi^0 \, dV - \int \varrho \, dV. \tag{27}$$

After circumscribing a sphere of radius $r > r_0$ about the charge we obtain E_n flux in the form

$$\int E_n \, ds = 4\pi m_v^2 \int_0^r \varphi^0(r) \, r^2 \, dr - 4\pi g \tag{28}$$

If

$$\varphi^0 = \frac{g}{r} e^{-m_v r},$$

$$\int E_n \, ds = 4\pi g \left[m_v^2 \int_0^r e^{-m_v r} r \, dr - 1 \right]. \tag{29}$$

In electrodynamics the vector E_n flux has the same value on the sphere of any radius, while in mesodynamics decreases with increasing radius of the sphere. When the sphere radius tends to infinity, the r.h.s. of Equation (29) vanishes; the flux E_n is completely damped. If in mesodynamics the term 'strength lines' is allowed then in the case of vector meson field the strength lines do not end, as in electrodynamics, in charges of opposite signes. Instead, they are damped in an original manner, by the 'field charge', if it is possible to say so, which is realized by the field potential, namely by the term $4\pi m_v^2 \int_0^r \varphi^0(r) r^2 \, dr$ playing the role of such a charge opposite in sign. This 'charge' is distributed over the whole space.

The sphere of finite radius r is crossed by a nonzero vector flux – this is the 'hair' of the vector meson field. In this case it is useful to consider (as we have agreed earlier) collapse of small masses, whose gravitational radius is much smaller than $\hbar/m_v c \sim$ $\sim 10^{-13}$ cm. The possibility of principle of collapse of small masses was discussed by Hawking (1970) and Zel'dovich (1962), An arbitrary small mass of a neutral matter can be brought into collapsing state by means of a huge pressure.

If the problem is to 'organize' a collapse of, e.g. one gram of neutrons then when neutrons are localized in a domain far smaller than $\hbar/m_v c$ the numerical value of the second term in Equation (28), 'field charge', becomes negligible. When the domain of localization is still larger than the gravitational radius of the system ($r_{gr} \sim 10^{-28}$ cm) by many orders of magnitude, when the gravitational forces may still be neglected the generalized Gauss theorem is of the form

$$\int E_n \, ds \cong -4\pi g. \tag{30}$$

In other words, there arises a purely electrodynamic analog of the Gauss theorem with all its consequences, in particular, and with respect to the baryon field of a system of neutrons, i.e. the vector meson field. If such a system was brought into a collapsing state then, in virtus of (30), this black hole would possess baryon 'hair'.

The assertion that outside a black hole the vector flux E_n vanishes implies that inside the black hole this flux vanishes too when approaching the surface of the event horizon. This means that the 'field charge' integrated over the internal space of the black hole increases in some way to the extent that it becomes able of compensating the disappearance of the potential which has occured outside the black hole. We cannot let the flux E_n vanish near the Schwarzschild sphere without an appropriate increase of the potential inside the black hole.

If we consider the situation with the flux in question in a closed world then it is easy to check that the potential (25) provides vanishing of the flux E_n on the 'boundary' for $\chi = \pi$. This possibility is explained by the fact that inside the closed world a term with increasing exponential is added to the expression for the potential. At this expense the density of the 'field charge' increases to the extent needed for the baryon charge to be compensated. *In this sense the total 'baryon charge' (r.h.s. of Equation (27)) in a closed world is zero. In this sense the closed world is also neutral in the baryon charge too.* Due to the change of the potential 'hair' of the vector meson field goes into the closed world.

The problem arises as to what is the reason for which the integral of the vector meson field potential inside a black hole increases when this object emerges in the process of collapse. The fact is that the boundary conditions at the euclidean infinity for a collapsing system are known to remain unchanged.

In other words, it is impossible to make the external field of a black hole only vanish. We should, figuratively speaking, 'drive' it inside the black hole so that to increase the integral $m_v^2 \int \varphi^0 \, dV$ inside the black hole up to a value which would compensate the total baryon charge (9).

In any case it is still unclear how the generalized Gauss theorem is fulfilled in the process of collapse. It is impossible for the time being to assert that black holes have no external baryon field.

On the other hand, the evidence that black holes have no external vector meson field under definite initial assumptions seems also to be convincing. The initial assumptions for this proof seem also to be quite natural: it is assumed that the weak meson vector field does not change the Schwarzschild metric, that there exists an event horizon and that the potential is finite at the event horizon.

The problem would be completely resolved if we succeeded in finding matchable internal and external solutions. Unfortunately, we have not yet derived an external static solution of the Nordström-Reissner type for the short-range, i.e. meson vector field. Until this programme is realized the above questions cannot be answered. An exact solution of the problem can give rise to some surprises. The example of the scalar field is, in this case, the most instructive one. In the case of the scalar field it is surprising that the event horizon is absent, and in those exceptional cases when it exists the field potential on the Schwarzschild sphere is found to be divergent.

2. The Scalar Long-Range Field

For the scalar field a problem similar to the Nordström-Reissner one was solved by Fischer (1948) a quarter of a century ago. Fischer obtained the metric in the form

$$ds^2 = \left(\frac{Z-Z_0}{Z+Z_1}\right)^p dt^2 - \frac{r^2}{Z^2}\left(\frac{Z-Z_0}{Z+Z_1}\right)^p dr^2 - r^2(d\theta^2 + \sin^2\theta \, d\varphi^2), \tag{31}$$

here

$$Z_{0,1} = (\varkappa^2 m^2 + \varkappa G^2)^{1/2} \mp \varkappa m, \tag{32}$$

G is the scalar charge,

$$p = \varkappa m(\varkappa^2 m^2 + \varkappa G^2)^{-1/2},$$

$$Z(r) = re^{(\nu-\lambda)/2} \underset{r\to\infty}{\to} r, \quad g_{00} = e^\nu, \quad g_{11} = -e^\lambda, \tag{33}$$

and

$$(Z - Z_0)^{1-p}(Z + Z_1)^{1+p} = r^2. \tag{34}$$

The calculations of Fischer were independently obtained by Janis et al. (1968). By a simple transformation the metric of the latter authors transforms into the metric (31). According to (31) g_{11} nowhere becomes infinite. Both g_{11} and g_{00} tend to zero, at $Z \to Z_0$.

Both papers contain some errors in the analysis of the asymptotic behaviour of the metric, but these errors do not concern the form of the linear element (31) which is calculated correctly. The metric (31) privides evidence that in the case of the long-range scalar field the horizon is absent, the Schwarzschild sphere is absent. In this case an object like the black hole is impossible.

It is remarkable that the event horizon is absent for any weak scalar field. But it is interesting that the transition to the limit (for a scalar charge $G = 0$) transforms finally the metric (31) to the Schwarzschild metric although not continuously.

In fact, for $G = 0$,

$$\begin{aligned} p &= 1, \quad Z_0 = 0, \\ Z_1 &= 2\varkappa m, \quad Z + 2\varkappa m = r. \end{aligned} \tag{35}$$

Inserting the values for $G = 0$ in the expression (31), we obtain the Schwarzschild metric. The metric (31) at $r \to 0$ and $G \to 0$ behaves in a nonanalytic way: Z as $r \to 0$, tends to zero too, but at the very limit $(r = 0)$ according to (35), Z assumes by jump the value (see Figure 1)

$$Z = -2\varkappa m. \tag{36}$$

If the external metric (31) is considered as valid for arbitrary small distances, then at $r \to 0$ there arises the case of a bare singularity with all serious consequences following from it.

Such a behaviour of the metric for small r testifies rather in favour of the invalidity of the metric (31) in vacuum for any small r.

It is appropriate to give certain simple considerations which stress particular features of the systems charged by the scalar field source. Let the bare mass of a matter distributed over a domain of radius r_0 be M_0 and the total scalar charge be given by G. Generalizing to this case the well-known relation (Arnowitt et al., 1960) for the total mass we get

$$M_{\text{tot}} = M_0 - \frac{\varkappa M_{\text{tot}}^2}{2c^2 r_0} - \frac{G^2}{2c^2 r_0}, \quad \text{or} \tag{37}$$

$$M_{\text{tot}} = -\frac{2r_0 c^2}{\varkappa} + \left(\frac{4r_0^2 c^4}{\varkappa^2} + \frac{2r_0 c^2 M_0}{\varkappa} - \frac{G^2}{\varkappa}\right)^{1/2}. \tag{38}$$

According to (38), the total mass of the system vanishes for

$$r_{\min} = \frac{G^2}{2M_0 c^2}. \tag{39}$$

The system in question cannot be localized in the domain $r < r_{\min}$. Here we meet one of the numerous peculiar properties of the scalar field.

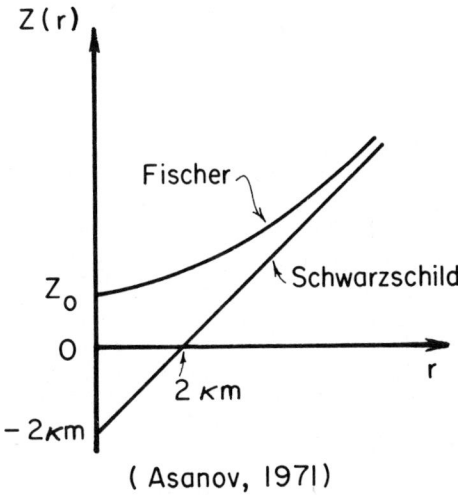

(Asanov, 1971)

Fig. 1. Fischer's solution for a point source of massless scalar field and the corresponding Schwarzschild solution.

From the above considerations concerning the metric (31) and the specific properties of the scalar field we draw the conclusion that the problem of the scalar field in the process of collapse (let us be careful) is still waiting for its solution. All the aspects of this problem will are clarified when one finds the internal solution of a collapsing system (taking into account the effect of the scalar field on the metric) and the external solution matched with it, and, which is essential, *outside the framework of perturbation theory*.

This problem is essentially nonstatic. It becomes more complicated by that, contrary to electrodynamics, here the central symmetric motion of matter can give rise to a monopole emission of a scalar field which alters the mass of the system. Above we were dealing with the scalar field, more exactly, with the traditional equation of the scalar field

$$\nabla_\sigma \nabla^\sigma U = 4\pi j.$$

But there exists another form for the scalar field equation (Penrose, 1963; Chernikov and Tagirov, 1968)

$$\nabla_\sigma \nabla^\sigma U + \frac{R}{\sigma} U = 4\pi j, \tag{40}$$

where R is the scalar curvature.

The problem is as to whether certain surprising results of the foregoing consideration may be due to the unfortunate form of the scalar equation or not. Sometimes authors stress that the conformal invariant Equation (40) has some advantages. However, the analysis of the problem with recourse to Equation (40) makes the situation more critical.

In this case the event horizon is also absent (Bocharova *et al.*, 1971).

But in a particular case for

$$G^2 = 3\varkappa m^2 \tag{41}$$

there arises the event horizon (the Schwarzschild sphere). There arises an external metric of the type (18):

$$g_{00} = e^\nu = e^{-\lambda} = \left(1 - \frac{a}{r}\right)^2,$$

$$a = \varkappa m = \left(\frac{\varkappa G^2}{3}\right)^{1/2}. \tag{42}$$

However, contrary to the electrostatic case, in this metric, the scalar potential on the event horizon surface becomes infinite:

$$U = -\frac{G}{r-a}. \tag{43}$$

In this case we have explicitly a couner-example which provides direct evidence that the black hole possesses an external scalar field.

As far as at the event horizon the potential U becomes infinite then the theorem that the field should vanish outside the black hole is inapplicable to the present case (Chase, 1970): it is essentially associated with the assumption about the finiteness of the potential on the Schwarzschild sphere.

We should bear in mind that the scalar curvature R in Equation (40) for the given case (in vacuum) vanishes. The equation for the scalar potential takes on the usual form. This fact gives, however, no grounds for concluding that a conformal invariant case of the theory must not differ from the ordinary one.

The matter is that we should consider the system of equations as a whole, while the Einstein equations have in this case, essential differences. So, one of the Einstein equations, in the writing $G^\nu_\mu = -8\pi\varkappa T^\nu_\mu$ is of the form

$$\left(1 - \frac{\varkappa U}{3}\right)\left(\frac{1}{r^2} - \frac{\lambda'}{r} - \frac{e^{-\lambda}}{r^2}\right) = -\frac{\varkappa}{3} U'^2 - \varkappa v'(U^2)'.$$

While, in a nonconformal invariant case:

$$\frac{1}{r^2} - \frac{\lambda'}{r} - \frac{e^{-\lambda}}{r^2} = -\varkappa U'^2.$$

The difference between these equations is a consequence of different expressions for the energy-momentum tensor T_μ^ν.

Attempts to consider the problem in a consistent way, find the appropriate internal and external solutions for a nonstatic metric, were made by Price (1971). Unfortunately these attempts were made in the framework of perturbation theory, under the assumption that it is possible to disregard the effect of the scalar field on the metric. If the metric (31) holds then this consideration is invalid. When using perturbation theory we should bear in mind that the peculiarity of the metric (31) consists in that the event horizon disappears for any value of the charge of the scalar field, for the exception of the limiting value $G=0$. However, using a particular example, the author comes to the conclusion that any catastrophic values of the scalar potentials do not arise at the horizon. But one should bear in mind that the properties of the scalar field are very perculiar (Dicke, 1964), and the particular example considered by the author is not free of objections (Markov, 1972). Besides, there is an explicit counter-example (42). This example gives also evidence for the invalidity of perturbation theory since any small violation of equality (41) results in vanishing of the horizon.

The instability of this solution recalls the instability of the Papapetrou solution. In fact, any small violation of the rigorous equality of gravitational attraction and electrostatic repulsion leads to a collapse. But the main point is that there is an explicit counter-example (43).

The most essential critical remark that can be made concerning the metric (31) is the following. It is doubtful whether the consideration of a purely static problem in the collapse of a matter charged by the scalar field sources is valid. The possibility of a monopole radiation of the scalar field can just change essentially the situation.

It may be noticed that the consideration of long-range scalar forces is of purely abstract interest, since it is very likely that this kind of forces does not exist in nature. May be, of more importance are the short-range scalar fields, scalar meson fields. Recently, these fields have been discussed in elementary particle theory and are of fundamental importance when attempting to construct a unified theory of weak and electromagnetic interactions (Weinberg, 1967). But the monopole radiation of scalar mesons in the process of collapse can be arbitrarily strongly suppressed by the large mass of the quanta of this field, and thus in this case the reproach for ignoring monopole radiation is cancelled. Moreover, the Price considerations about the role of the effective potential barrier are no longer applicable to this 'short-wave' field.

The finding of the external metric of a spherical symmetric source of the meson scalar field still encounter unsolved difficulties.

Recently Asanov (1973) obtained a numerical solution for the system of the Einstein equations and the Klein equation

$$\left(\nabla_\sigma \nabla^\sigma + \frac{\mu^2 c^2}{\hbar^2}\right) U = 4\pi j \tag{44}$$

for the spherical symmetrical case we are interested in.

The numerical solution was found for the following values of the parameters

$$\varkappa G^2 = \varkappa^2 m^2, \qquad \mu = \frac{1}{\varkappa m} = \frac{1}{\varkappa^{1/2} G}. \tag{45}$$

The results of calculations are given in Figure 2. The function e^λ of its values at infinity $(e^{-\lambda} \approx 1 - (2\varkappa m/r))$ increases smoothly and reaches its maximum $(e^\lambda \approx 9.5)$

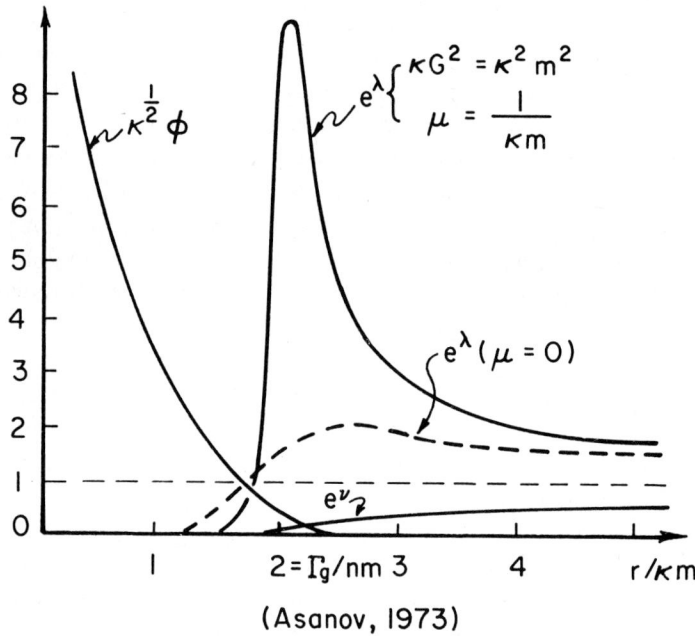

(Asanov, 1973)

Fig. 2. $g_{00} = e^\nu$ and $g_{\parallel} = e^\lambda$ in the case of scalar field of a point source of mass μ. e^ν and e^λ are equal to zero at $r \to 0$.

near 'the gravitational radius' $r_{gr} = 2\varkappa m$. Then e^λ falls tending smoothly to zero for $r = 0$. The function e^λ is equal to unity for $r \approx 1.8\ \varkappa m$. The function e^ν is monotonous, it is equal to zero for $r = 0$ and to unity at the spacial infinity.

If we assume that at present the existence of the black hole is confirmed by experimental data, and that on the other hand, the above theoretical results concerning the specific properties of the scalar field, incompatible with the existence of black holes, are valid, then we should conclude that the presence of black holes would testify to the absence of scalar mesons in nature. This would mean that the well-known theoretical attempts trying to unify weak and electromagnetic interactions on the basis of symmetry breaking are groundless. And, on the contrary, if the presence of scalar mesons is confirmed by experiment then the objects which at present are thought of as black holes will be interpreted in some other manner. In the theoretical aspect we should be led to the conclusion that in real situations there never arise $T_\pm$ domains the possibility of which we have already got accustomed to, and the metrics of Figure 3 becomes much simpler. At the same time, there arise objects close to bare singularities with all the consequences resulting from this. In general,

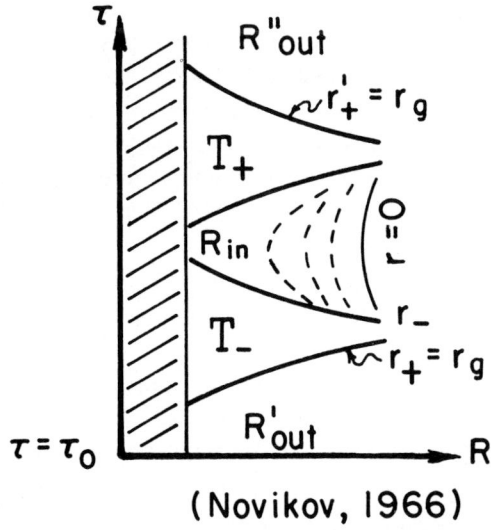

Fig. 3. The evolution of a charged sphere after collapse behind the Schwarzschild surface without taking into account the process of pair creation.

the 'price' of the proof of the validity or invalidity of the static metric (31) up to distances smaller than the gravitational radius is very high, and we should be very careful in estimating all arguments for and against. The numerical solution for, e.g. a model with scalar and electrostatic field sources, shows that up to $r = 0.9 \varkappa m$, where $e^\lambda > 1$ and it is possible to match the internal and external solutions, the horizon is absent. (Asanov, 1972.)

3. The Neutrino Field

The situation with neutrino forces is far more complicated. In fact, if there is a direct interaction of the type $(ev)(ev)$ then the system consisting e.g. of hydrogen should induce in the surrounding space a neutrino-antineutrino field with potential (Tamm, 1934):

$$B \sim \frac{1}{r^5}.$$

There is a direct calculation (Hartle, 1971) for the vector-pseudovector case where it is shown that when neutrino field sources come nearer the Schwarzschild sphere the neutrino field in the external space tends to zero. But the calculation of the scalar case by this method does not result in the disappearance of neutrino hair outside the black hole (Berezin, 1973). Moreover, for a point of neutrino forces of the vector case localized at the point $\chi = 0$ of a closed world, as in electrostatics, there arises a mirror image of the sorce for $\chi = \pi$ (Berezin).

In other words, one would think that the appropriate neutrino hair should not disappear. On the other hand, it would seem that Hartle (1971) gives for the Kerr metric a general proof of the absence of the external neutrino field of holes independently of the type of forces. This proof is also valid for the Schwarzschild metric.

If in further investigation we do not find errors in the calculations for the scalar case then the general consideration of Hartle contains some defects.

Unfortunately, the field under discussion is not the solution of some equation of the Maxwell type, and the conservation law for leptons is not associated with any analog of the Gauss theorem. It appears that in the case of the neutrino field, as in the foregoing cases of meson fields, the ultimate solution for the behaviour of these fields outside black holes arises in the process of finding matchable internal and external solutions for the systems in questions.

The foregoing consideration of the external metrics of black holes with different charges should be supplemented with one general, and in our opinion, essential remark.

In fact, as yet studies have not been made of the inverse process, process of formation of hair (e.g. vector meson field, scalar field) which inevitably occurs in the process of anticollapse, when matter goes out of the Schwarzschild sphere.

4. Black Holes as Intermediate States in Elementary Particle Theory

We are not aware of whether black holes exist in nature. But there some grounds to suppose that these objects and the corresponding notions may turn out to be essential in elementary particle theory.

In fact, in contemporary theory of elementary particles the calculation of proper masses leads to divergent values. This is due to the fact that the intermediate states in these calculations may have energy, and consequently, mass of arbitrary large values.

In contemporary elementary particle theory a striking violation of logic became historically legitimate: one introduces in consideration intermediate states with arbitrary large masses and, at the same time, entirely ignores their gravitational effects. It is of special importance that these masses must be localized in a very small domain so that there arises the possibility of huge gravitational mass defects capable of changing in a cardinal manner the energy spectrum of intermediate states.

In fact, if a particle in the intermediate state emits a quantum of mass $m = \hbar v/c^2$ then, according to the Heisenberg relation*, this mass is localized in the domain

$$l \sim \frac{\hbar}{mc}, \quad \text{or} \quad m = \frac{\hbar}{lc}. \tag{46}$$

---

* When emitting a quantum of an energy $E = hv$ a particle, using the Fermi terminology, 'borrows' an energy $E = mc^2$. According to the uncertainty relation, the time of 'borrowing' may not be longer than $\hbar/mc^2$. During this period the emitted quantum cannot go away from the particle not far than at a distance $\sim \hbar/mc$.

When a mass of the order of

$$m \sim \left(\frac{\hbar c}{\varkappa}\right)^{1/2}$$

appears in the intermediate state, then the gravitational radius of this system is

$$r_{\text{gr}} = \frac{2\varkappa m}{c^2} = \frac{2(\varkappa \hbar c)^{1/2}}{c^2}. \tag{47}$$

On the other hand, with this mass the dimension of the domain in which, according to (46), the mass is localized

$$l \sim \frac{\hbar}{mc} \sim \frac{(\hbar c \varkappa)^{1/2}}{c^2} \tag{48}$$

coincides with the gravitational radius of the object in this state. With further increasing intermediate state energy the gravitational radius should have increased. But, on the other hand, the domain of localization of the intermediate state energy should, according to the Heisenberg relation, have decreased and for $m > (\hbar c/\varkappa)^{1/2}$ should have became smaller than the gravitational radius.

If such a situation arose in the range of applicability of classical physics we would say that we are dealing with a system whose mass is behind the gravitational Schwarzschild sphere. In other words, we would imply a system in the collapsing state. This might be either the state of black hole of the first kind or rather the state of systems with semi-closed matric if the bare mass of the intermediate state decreases strongly due to gravitational defect. From this viewpoint it would seem that the states of semi-closed systems or the states of black holes must belong to the complete set of states that can arise spontaneously in these cases. Moreover, energetically these states are the lowest states. Although at present we do not know to what extent our understanding of the metric remains valid in this state. However, in modern theory one got accustomed to use the euclidean metric at arbitrary small distances from the point particle. If black holes have really no hair then in the spectrum of the mediate states such states must in a cardinal manner, affect the results of calculations, since in these states there are no longer interactions with the given fields.

Appendix 1

(The role of the vector meson field in a distant stage of collapse of stars).

Novikov (1966) considered the evolution of a charged sphere after collapse behind the Schwarzschild sphere. The spacetime picture of this evolution is given by Novikov in Figure 3 in the Cruscal type coordinates. The region occupied by matter is shaded.

In the course of contraction its boundary crosses the Schwarzschild sphere. For a charged sphere of mass M and charge ε

$$r_{\text{gr}} = r_+ = \frac{\varkappa M}{c^2}\left[1 + \left(1 - \frac{\varepsilon^2}{\varkappa M^2}\right)^{1/2}\right],$$

where $\varkappa$ is the gravitational constant, c – the light velocity. From the external domain (R'_{out}) the system falls in a contracting T_- – domain.

For the sizes

$$r_- = \frac{\varkappa M}{c^2}\left[1 - \left(1 - \frac{\varepsilon^2}{\varkappa M^2}\right)^{1/2}\right]$$

contraction can be changed by expansion. This occurs in the internal (R_{in}) domain. After passing the expanding T_+ domain the sphere boundary crosses again the Schwarzschild sphere and goes out in the external (R''_{out}) domain (Figure 3). However, as is seen from the figure, the domain R''_{out} is not the same in which the contraction of the sphere occurs. The R''_{out} domain lies in an absolute future with respect to R'_{out}. The peculiarity of the situation consists in that the behaviour of the system after crossing the surfaces r_- is not defined by the initial conditions in the R'_{out} space. Thus, in mechanics there first arises the possibility for the Laplace determinism to be violated*.

As it happens always in these cases, congenial scientific conservatism searches naturally for possibilities of solving the problem in some other way, namely in the framework of usual regularities. It is a cruel necessity alone that can affect in a cardinal way the existing viewpoints.

In fact, it seems that there is such a natural possibility. The matter is that in the process of contraction of a charged sphere there arise outside it electric fields of a so high intensity that *a purely classical consideration of the problem neglecting the production of different kinds of charged particle pairs in these fields is quite inadmissible.*

In fact, for $\varepsilon^2 \ll \varkappa M^2$, due the process of contraction, a charged sphere on the boundary of the T_- domain acquires dimensions $r_- \sim \varepsilon^2/2Mc^2$.

The electric potential on the surface of this sphere should have taken the value:

$$V \sim \frac{\varepsilon e}{r_-} = Mc^2 \frac{e}{\varepsilon}, \quad \text{or} \quad eE = \frac{\varepsilon e}{r_-^2} = \frac{eM^2c^4}{\varepsilon^3}.$$

For $M = M_\odot \sim 10^{34}$ g, $\varepsilon < 10^{20}$ e.

In other words, the electrostatic field energy, localized outside a sphere of radius r, turns out to be equal to the total energy of the system. However, in reality, such a potential cannot be realized due to electron-positron pairs production effect. Electron-positron pairs are produced long in advance until the charged sphere reduces to dimensions r_-.

Thus, the situation arises when because of production of new particles we are

---

* The reference of Penrose (1968) to the fact that in quantum theory we have got accustomed to indeterminism is a misunderstanding: the behaviour of the wave functions is completely defined by the initial data.

deprived of the convenient *comoving coordinate system* for describing the process of collapse and, moreover, the classic space-time description losses its meaning when a pair emerges, electrons and positrons may be spaced from one another by an essential space-like interval (Zel'dovich, 1971). It is possible that electrons should not reach the surface r'_- if, (for instance) in T_-, and R_{in} domains will appear a charge of opposite sign (Novikov, 1970) or the all picture may be essential the other.

It should be noticed that the inclusion of the electric charge in the system is a very artificial method and bears no direct relation to a really collapsing star. However, the really collapsing star is immediately associated with the meson vector field.

The vector meson field is a short-range one, therefore until a collapsing system turns out to be contracted up to dimensions of the order $\hbar/m_v c$ (m_v – the vector meson mass), this field may not be of importance in the process of collapse. However, with further contraction ($r < \hbar/m_v c$) the vector field becomes practically a long-range one, but at the same time it is stronger than the electromagnetic field. The meson field is capable of producing nucleon-antinucleon pairs.

The model of the electrically charged sphere has the shortcoming that the electric charges are, in this model, a mechanical admixture to neutral matter. Strictly speaking, the whole previous consideration of collapse of electrically charged matter is meaningful when each particle of this matter is electrically charged.

In the case of an electrically neutral star, each of its nucleons is a carrier of the meson vector field charge.

It should also be added that the collapsing mass of a star of the order of the solar mass has a density $\varrho \sim 10^{72}$ g cm^{-3} at dimensions $\sim \hbar/m_v c \sim 10^{-13}$ cm which is by 20 orders of magnitude smaller than the critical 'quantum' density at which, as one hopes without any definite grounds, collapse of a star can change by anticollapse. Thus, there are serious grounds to assume that the appearance of R_{in} and R''_{out} in Figure 3 and the violation of the Laplace determinism is a consequence of an abstract approach to the process.

It is remarkable that the critical density $\varrho_q \sim 10^{93}$ g cm^{-3} would be reached for collapsing masses

$$m \sim \varrho_q \left(\frac{\hbar}{m_v c}\right)^3 \sim 10^{55} \text{ g}$$

i.e. of the order of the mass of the whole Universe.

Appendix 2

A possible appearance of a semi-closed metric in the process of anticollapse (in particular, white holes) was considered by Frolov (1973).

Frolov considered a more general case of motion of a massive charged, radiating spherically symmetrical shell. The external metric of a radiating system found by Vaidya (1971, 1953) was used in his paper. The equations of motion of the shell are derived following Israel (1967, 1968). A detailed analysis is given for the case, when radiation occurs during a short period of the proper time. In this case the connection is found of the parameters of the system before and after radiation or absorption of both energy and charges of the system.

The effect of these processes on the total mass of the system for different types of the shell (anticollapse, collapse, usual motion) is analysed. It is shown that in the case of an open system with small charge, a transition to semiclosed state can occur due to energy radiation. If the radiation transfers the charge, then the final states can have the friedmons metric.

References

Ambartsumian, V. A.: 1962, *Voprosy Kosmologii* **8**, 3.
Arnowitt, R., Deser, S., and Misner, C.: 1960, *Phys. Rev.* **120**, 313.
Asanov, R.: 1972, preprint P2-6564, Dubna.
Asanov, R.: 1973, preprint P2-7230, Dubna.
Bardeen, J. M.: 1968, *Bull. Am. Phys. Soc.* **13**, 41.
Bekenstein, J.: 1972, *Phys. Rev. Letters* **18**, 452.
Bekenstein, J.: 1973, *Phys. Rev.* **D7**, 949.
Berezin, V.A. and Markov, M. A.: 1970, *Teor. Mat. Fiz.* **3**, 161.
Blokhintsev, D.I.: 1960, *Nuovo Cimento* **16**, 382.
Carter, B.: 1966, *Phys. Letters* **21**, 423.
Chase, J.: 1972, *Commun. Math. Phys.* **19**, 276.
Chernikov, N.A. and Tagirov, E. A.: 1968, *Ann. Inst. Poincaré* **9**, 1507.
De la Cruz and Israel: 1967, *Nature* **216**, 148, 312.
Dicke, R.: 1964, in Hong-Yee Chiu and W. Hoffmann (eds.), *Gravitation and Relativity*, W. A. Benjamin Inc., New-York-Amsterdam.
Fischer, I.: 1948, *JETP* **18**, 636.
Frolov, V. P.: 1973, preprint Lebedev Inst., Moskow.
Hartle, J.: 1971a, *Phys. Rev.* **D3**, 2938.
Hartle, J.: 1971b, preprint.
Hawking, S. W.: 1970, *Monthly Notices Roy. Astron. Soc.* **152**, 75.
Hawking, S. W.: 1971, *Phys. Rev. Letters* **26**, 1344.
Hoyle, F., Fowler, W. A., Burbridge, G. R., and Burbridge, E. M.: 1965, *Quasi-Stellar Sources and Gravitational Collapse*, University of Chicago Press.
Israel, W.: 1966, *Nuovo Cimento* **44B**, 1.
Israel, W.: 1967, *Nuovo Cimento* **48B**, 463.
Janis, A. I., Newmann, E. T., and Winicour, J.: 1968, *Phys. Rev.* **176**, 1507.
Klein, O.: 1961, *Werner Heisenberg und die Physik Unserer Zeit*, Braunschweig.
Markov, M. A.: 1966, *JETP* **51**, 878.
Markov, M. A.: 1970, *Ann. Phys.* **59**, 109.
Markov, M. A.: 1971, *Cosmology and Elementary Particles* (Lecture Notes), Trieste IC/71/33 Part I and II.
Markov, M. A.: 1972, preprint E2-6831, Dubna.
Markov, M. A. and Frolov, V. P.: 1970, *Teor. Mat Fiz.* **3**, N1, 3.
Markov, M. A. and Frolov, V. P.: 1972, *Teor. Mat. Fiz.* **13**, 41.
Markov, M. A. and Frolov, V. P.: 1973, *Teor. Mat. Fiz.*
Ne'eman, Y.: 1965, *Appl. J.* **141**, 1303.
Novikov, I. D.: 1962, *Vestn. Mosk. Gos. Univ. Ser. 3*, N5.
Novikov, I. D.: 1964, *Astron. J.* **41**, 1975.
Novikov, I. D.: 1966, *Pisma JETP* **3**, 223.
Novikov, I. D.: 1970, *JETP* **59**, 262
Papapetrou, A.: 1945, *Proc. Roy. Phys. Acad.* **L1**, Sec.A., 191.
Penrose, R.: 1968, *Structure of Space-Time*, W. A. Benjamin inc., New-York-Amsterdam.
Regge, T.: 1958, *Nuovo Cimento* **7**, 215.
Regge, T. and Wheeler, J. A.: 1957, *Phys. Rev.* **108**, 1963.

Sakharov: 1970, preprint N7, Inst. Prikladnoy Mat. AC.N.C.C.C.R.
Tamm, I.: 1934, *Nature* **134**, 1010.
Teitelboim, C.: 1972, *Lettere al. Nuovo Cimento* **3**, 326, 397.
Thorne, K.: 1971, preprint OAP-236.
Vaidya, P. C.: 1951, *Phys. Rev.* **83**, 10.
Vaidya, P. C.: 1953, *Nature* **171**, 260.
Wheeler, J. A.: 1971,
Weinberg, S.: 1967, *Phys. Rev. Letters* **19**, 1264.
Zel'dovich, Ya. B.: 1962a, *Zh. Exp. Teor. Fiz.* **42**, 641.
Zel'dovich, Ya. B.: 1962b, *Zh. Exp. Teor. Fiz.* **43**, 1937.
Zel'dovich, Ya. B.: 1971, preprint N1, Institute of Applied Mathematics of the Acad. Nauk of the U.S.S.R.

PROPERTIES OF BLACK HOLES RELEVANT TO THEIR OBSERVATION*

JAMES M. BARDEEN

Yale University, U.S.A.

Abstract. Black holes are very small objects by astronomical standards, so in many circumstances they interact with their surroundings like a Newtonian mass point. However, if black holes are present in X-ray binary systems, the X-rays emitted in the inner part of the accretion disk probe the highly curved spacetime geometry near the horizon, particularly if the black hole is rapidly rotating. Some of the properties of circular orbits near the black hole are quite sensitive to the amount of angular momentum. The relativistic corrections remove a Newtonian degeneracy between several of the characteristic frequencies associated with perturbations of the circular orbits.

Hot spots in the inner part of the disk can produce dramatic fluctuations in intensity, since the frequency shifts of photons emitted by a given point on the disk are strongly time-dependent. The bending of the photon trajectories by the strong gravitational field can drastically affect the energy balance of the disk; much of the radiation emitted by the inner part of the disk is reabsorbed. The dragging of inertial frames by the angular momentum of the black hole can have striking consequences for the structure of the disk at quite large radii if the angular momentum of the accreting matter is not in the same direction as the angular momentum of the black hole.

Dynamic perturbations of black holes are now being intensively studied to see if there are any surprising physical effects associated with the rotation of the black hole. Unfortunately, though quite interesting methods of extracting energy from the black hole exist in principle most of them are unlikely to be realized to an important extent in the real astrophysical world.

1. Introduction

To a relativity theorist the most interesting observational aspects of black holes are those that reveal something of their intrinsic general relativistic nature. Ideally, one would like some observational confirmation of the Israel-Carter conjecture that the Kerr family of solutions to the Einstein equations includes all physically acceptable vacuum black hole geometries. The current indications that black holes probably do exist in certain X-ray binaries raise the hope that in the not-too-distant future we will be able to test strong field predictions of general relativity in a relatively clean way, without the uncertainties of the equation of state and radiation mechanisms that complicate interpretations of pulsar observations.

Therefore, I will concentrate in this lecture on the properties of black holes which strike me as most relevant to such strong field observations. Peebles (1972) has given an excellent review of how black holes might be detected through their long-range Newtonian gravitational field in globular clusters, galactic nuclei, etc. The detailed astrophysics of accretion into black holes will be reviewed at this Symposium by Sunyaev.

In order to observe strong-field effects some source of radiation must be present near, but outside the horizon of the black hole. The most likely source of the radiation is accreting matter flowing in some more or less continuous way into the black hole.

---

* Supported in part by the National Science Foundation (U.S.A.).

Another, more remote possibility is a small, compact source of radiation (a 'star') orbiting near the black hole. Finally, the black hole may amplify an external wave disturbance or radiation may be emitted as the black hole settles down after being formed in gravitational collapse.

I assume that the metric describing the black hole is one of the Kerr metrics (Kerr, 1963). Theorems of Israel (1967), Carter (1971), and Hawking (Hawking and Ellis, 1973) establish almost conclusively that the Kerr metrics are the only stationary vacuum solutions of the Einstein equations which have regular event horizons and are asymptotically flat. The charged black holes, described by the Kerr-Newman metrics (Newman *et al.*, 1965) are of little astrophysical interest. There are few mechanisms of charge separation in gravitational collapse capable of producing a significant charge on a black hole, and any initial charge will rather quickly be neutralized, since astrophysical plasmas are good conductors.

The Kerr metrics are a two-parameter family, characterized by gravitational mass M and angular momentum J. Event horizons exist if and only if M and J satisfy $0 \leq cJ/GM^2 \leq 1$. It seems reasonable that centrifugal forces will prevent the complete collapse of any object with $a/M = cJ/GM^2 > 1$; if a black hole is formed, some matter carrying the excess angular momentum must be left behind.

What values of the parameter a/M can we reasonably expect for astrophysical black holes? This is an important question for the possibility of observing strong field effects. While if $a/M = 1$ matter can exist in stable circular orbits arbitrarily close (in some senses) to the horizon, the radius of the innermost stable circular orbit increases rapidly as a/M decreases from one, from $r/M = 1$ when $a/M = 1$, to $r/M \simeq 2$ when $a/M = 0.95$, to $r/M \simeq 4$ when $a/M = 0.5$, and to $r/M = 6$ when $a/M = 0$ (all in units with $G = c = 1$). The ergosphere, where energy can be extracted from the black hole by emission of radiation or energetic particles in appropriate directions, extends from the horizon at $r = r_+ = M + (M^2 - a^2)^{1/2}$ to $r = 2M$ in the equatorial plane.

Calculations of Bardeen and Wagoner (1971) suggest that a black hole with a/M infinitesimally close to one can be formed only by a quasi-stationary collapse which passes through a series of disk-like configurations. These disk-like configurations are probably unstable to a fragmentation process in which the central part of the disk, with $a/M < 1$ by a finite amount, collapses dynamically. Thus, while many objects in the Galaxy have $cJ/GM^2 \gtrsim 1$ (the Sun, for instance), it seems unlikely that black holes will form with a/M very close to one, though perhaps with a/M as large as 0.9 or so.

Accretion of matter following the formation of the black hole can increase a/M. Bardeen (1970) has shown that an accretion process in which matter falls into the black hole from the innermost stable circular orbit in the equatorial plane can make $a/M = 1$ after a finite amount of rest mass is accreted, provided that the binding energy released by the matter in getting into the innermost stable circular orbit all escapes to infinity. More realistic calculations, in which some radiation emitted by the accreting matter is captured by the black hole, suggest that accretion might be able to increase a/M to a limiting value of perhaps 0.998 (Thorne, 1974). However, the rate of accretion cannot be so great that the radiation pressure of the energy released by the accreting matter exceeds the gravitational attraction of the black hole. The Eddington limit restricts the characteristic time scale for accretion to modify the black hole parameters to be greater than about 10^8 yr. This sort of time scale is probably considerably longer than accretion can be maintained in a binary system composed of massive stars, but is not unreasonable for a large black hole in a galactic nucleus.

2. Geodesics in the Kerr Metric

Many observational aspects of black holes concern the properties of timelike and null geodesics. In the accretion disk models that have been suggested for galactic nuclei by Lynden-Bell (1969) and Lynden-Bell and Rees (1971) and elaborated for some of the X-ray binaries by Pringle and Rees (1972), Shakura and Sunyaev (1973), and Novikov and Thorne (1973) the matter to a good approximation follows circular geodesic orbits in the equatorial plane of the black hole. Such accretion disks seem to offer the best chance for tests of black hole theory in the near future.

Spherical accretion, with the black hole at rest in the interstellar medium, is discussed extensively by Zel'dovich and Novikov (1971) and the detailed observational consequences are examined by Shvartsman (1971) and by Shapiro (1973). Novikov and Thorne (1973) conclude that synchrotron radiation from such a black hole might be observable, but spherical accretion offers few good observational handles to probe detailed properties of the black hole. At least in some cases the radial flow of the gas is close to free fall.

In all cases the propagation of the electromagnetic radiation emitted by the matter is along null geodesics since all observable wavelengths are extremely small compared with the radius of the black hole.

Fortunately, as first shown by Carter (1968), the Hamilton-Jacobi equation for geodesics of the Kerr metric separates and the trajectories can be determined by quadratures. A complete set of constants of the motion for a test particle with momentum four-vector p_a are $E = -p_t$, the energy, $L_z = p_\phi$, the angular momentum about the axis of symmetry, $\mu = (-p_a p^a)^{1/2}$, the rest mass, and finally the separation constant

$$K = p_\theta^2 + (L_z - aE \sin^2\theta)^2/\sin^2\theta + a^2\mu^2 \cos^2\theta. \tag{1}$$

If $a = 0$, so the black hole is spherically symmetric, K is the square of the total angular momentum of the test particle. If $a \neq 0$ the total angular momentum is in general not a constant of the motion, and K is not the total angular momentum even at infinity where the total angular momentum is constant.

Properties of timelike geodesics have been discussed by Ruffini and Wheeler (1970) and by Wilkins (1972), among others. Simplified formulas governing trajectories in the equatorial plane are given in Bardeen et al. (1972).

One immediate result of Equation (1) is that all orbits in the equatorial plane have

$$Q \equiv \dot{K} - (L_z - aE)^2 = 0. \tag{2}$$

Also, if $E=\mu$, $L_z=0$, and $Q=0$, the trajectories are 'radial' in the sense θ is constant for all values of θ, though ϕ does vary. Such trajectories represent matter falling from rest at infinity with zero total angular momentum. As long as the flow is highly supersonic near the horizon calculations of 'spherical' accretion are not much more complicated in the Kerr geometry than they are in the spherically symmetric Schwarzschild geometry. However, a detailed calculation is necessary to see if there are any very striking observational consequences of the black hole's rotation in this case.

The rotation of the black hole does have important consequences for the case of disk accretion. If the angular momentum of the accreting material is in the same direction as the angular momentum of the black hole, the accretion disk will lie in the equatorial plane. The total gravitational energy released and the position of the inner edge of the accretion disk depend strongly on the value of a/M, particularly near $a/M=1$. The ratio E/μ for direct circular orbits in the equatorial plane of the black hole is plotted versus proper radial distance in Figure 1. The innermost stable

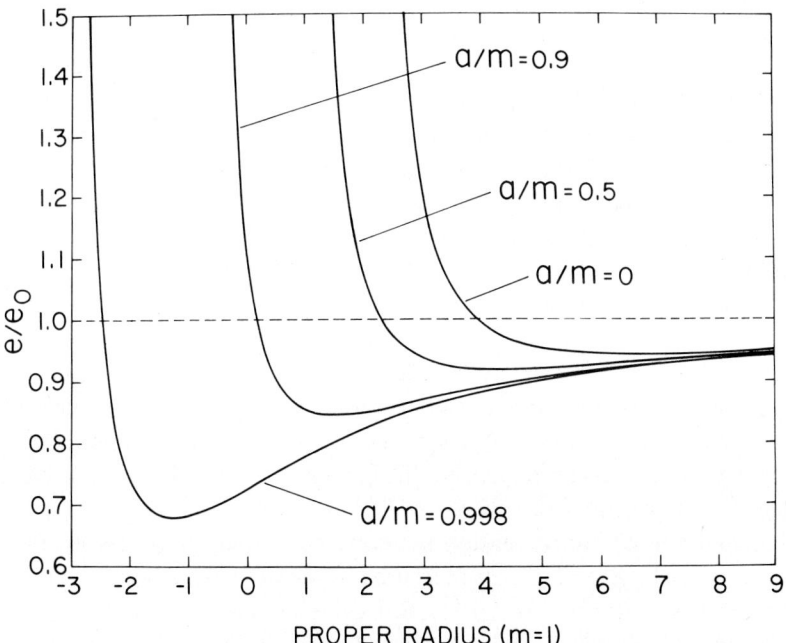

Fig. 1. The ratio of energy e to rest mass energy e_o as a function of proper radius for circular orbits in the equatorial plane of some Kerr black holes. In the text e and e_o are denoted by E and μ, respectively, and the mass m of the black hole is denoted by M. When $a/m=cJ/GM^2=1$ the ratio e/e_o reaches a minimum value of 0.58 at a proper radius of minus infinity.

circular orbit is at the minimum of E/μ and represents the inner edge of the accretion disk in a simple model. The total gravitational energy release per unit rest mass in the accretion disk is $1-(E/\mu)_{\min}$. While an often quoted number is $1-(E/\mu)_{\min}=0.42$ for $a/M=1$, even for $a/M=0.998$, $1-(E/\mu)_{\min}$ is only 0.32. When a/M is close to one the ratio of proper radius interval to coordinate radius interval is large near the horizon. The innermost stable circular orbit is at coordinate radius $r \simeq M + [4(1-a/M)]^{1/3}$; the exact value for $a/M = 0.998$ is $r/M = 1.237$.

If some way could be found to calibrate the mass accretion rate independent of the luminosity of the accretion disk, the efficiency of energy release would be a sensitive test for the value of a/M.

When a/M is close to one a sizable fraction of the gravitational energy release occurs inside the ergosphere, at $r < 2M$. However, this situation is not as favorable for observing the effects of the strong gravitational field as it might seem, since the turbulent viscosity in the disk transfers most of the energy out to somewhat larger radii before it is radiated (Novikov and Thorne, 1973).

Much of the hope for testing strong field predictions of general relativity lies in the possibility of observing local inhomogeneities or 'hot spots' in the inner part of the disk which would give rise to fluctuations in the X-ray output of the disk. Sunyaev (1972) has suggested that the periodic time-varying Doppler shift of the radiation emitted by such a hot spot would give a measure of the period of rotation of the disk. The period of rotation at the inner edge of the disk varies from $2\pi(6)^{3/2}M$ for $a/M = 0$ to $4\pi M$ for $a/M = 1$. If $M = 10\ M_\odot$ this is a variation in the minimum period from 5 ms to 0.6 ms. The shortest time variations presently observable are roughly 100 ms, but with bigger collecting areas on future X-ray satellites this may be reduced considerably. Since most of the energy flux comes from somewhat outside the inner edge of the disk, the dominant brightness variations might have periods several times larger than the numbers quoted above.

Other sources of fluctuations might be local oscillations of a region on the disk about its steady state. Some of these oscillations might be associated with the internal stresses in the disk, while others might be associated with perturbations in the orbits of the fluid elements. Some characteristic frequencies for the latter type of perturbation are plotted against coordinate radius, along with certain other physically important quantities, in Figure 2. The value of a/M is 0.998, but the main effect of changing a/M is to change the radius of the inner edge of the disk, rather than change the values of the quantities at a fixed r/M.

The angular velocity of revolution as seen from infinity is Ω. The quantities $K_\parallel$ and $K_\perp$ are epicyclic frequencies as measured by a comoving observer. If the circular geodesic orbit is perturbed, the perturbed orbit oscillates about the unperturbed orbit with the frequency $K_\parallel$ if the perturbed orbit lies in the equatorial plane and with the frequency $K_\perp$ if the perturbation is perpendicular to the equatorial plane. Through the equation of geodesic deviation, $K_\parallel$ is a measure of the tidal force acting in the radial direction in the comoving frame of the circular orbit, and $K_\perp$ is a measure of the tidal force acting in the θ-direction in the comoving frame. By the axial symmetry, there is no such tidal force in the ϕ-direction. The quantity $\bar{\Omega}$ is the locally-measured angular velocity of rotation of the comoving frame, which is tied to the symmetry directions of the local (tidal) gravitational field, with respect to a non-rotating (Fermi-Walker-transported) frame. This angular velocity determines the centrifugal forces acting in the comoving frame.

While Ω, $\bar{\Omega}$, and $K_\odot$ increase roughly as $r^{-3/2}$ all the way in to the inner edge of the disk, $K_\parallel$, by definition, is zero at the marginally stable circular orbit. All four quantities are equal in the limit $r \gg M$. The local rate of shear of the disk, σ, is related to $K_\parallel$ and Ω by

$$K_\parallel = [2\Omega(2\Omega - \sigma)]^{1/2}, \tag{3}$$

and the vorticity, twice the angular velocity of a fluid element about its own center of mass, is $2\Omega - \sigma$.

The velocity v of the fluid element with respect to the zero-angular-momentum frame is more directly relevant to the appearance of the disk, since it gives the angle a zero angular momentum photon makes with the ϕ-direction in the comoving frame.

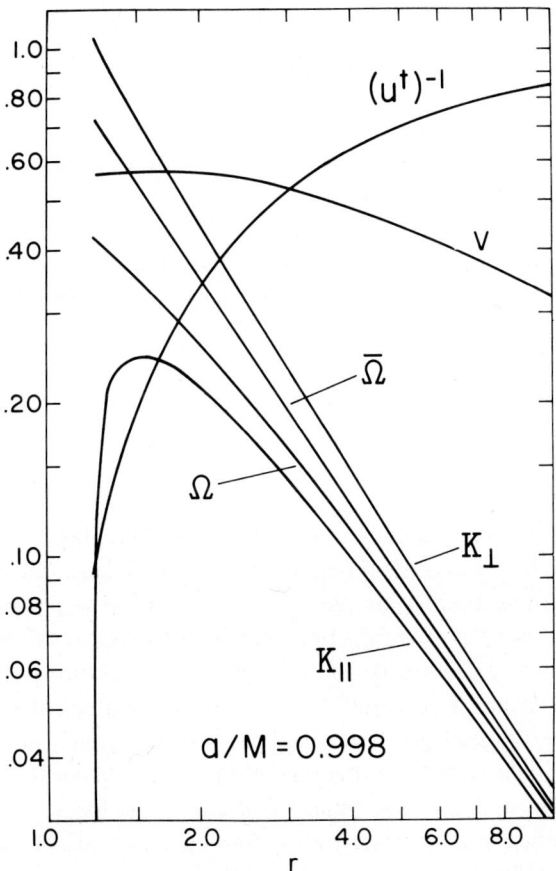

Fig. 2. Several physically important quantities characterizing circular orbits in the equatorial plane of a Kerr black hole with $a/M = 0.998$ are plotted against coordinate radius r. The units are such that $c = G = M = 1$. The angular velocity of revolution of a test particle about the black hole as seen from infinity, Ω, the angular velocity of rotation of the local tidal force field, $\bar{\Omega}$, the epicyclic frequencies in and perpendicular to the equatorial plane, $K_\parallel$ and $K_\perp$, are important for understanding the dynamics of accretion disks. The quantity $(u^t)^{-1}$ is the zero-angular-momentum redshift factor and v is the velocity of revolution with respect to a zero-angular-momentum observer.

The *local* frequencies are redshifted by the factor $(u^t)^{-1}$ as seen from infinity on a time-averaged basis. However, the instantaneous redshift factor varies from a little less than $(u^t)^{-1}$ to (for a small fraction of the orbital period) roughly one if the observer is within about 30° of the equatorial plane. Therefore, dynamic oscillations of a region on the disk are not completely frozen by the time dilation effect even if they take place near the inner edge of the disk for a/M close to one. A locally periodic oscillation at r/M close to one is not seen as a periodic oscillation from infinity. It comes alive for a fraction of the orbital period, when the local region is seen as relatively bright, and then tends to 'freeze' for the rest of the orbital period, when the local region is faint because its radiation suffers a large redshift. The very central part of a disk about an extreme Kerr hole will always be faint and time dilated when the observer is more than about 60° from the equatorial plane.

To get a complete picture of the appearance of an accretion disk and to understand the transfer of energy by radiation in the region near the black hole one must look at the null geodesics in some detail. A null geodesic trajectory can be characterized by two non-trivial parameters, $\lambda = L_z/E$ and $\eta = Q^{1/2}/E$. These parameters can be related to the direction cosines of the beam of radiation in the frame comoving with the matter at the point of emission, on the one hand, and to the impact parameters which describe the position of the image seen by a distant observer, on the other hand. For an observer at a polar angle θ_0 the impact parameter measuring the apparent perpendicular distance of the image from the symmetry axis is

$$\alpha = -\lambda/\sin\theta_0, \tag{4}$$

and the impact parameter measuring the apparent perpendicular distance of the image from the projection of the equatorial plane is

$$\beta = (\eta^2 + a^2 \cos^2\theta_0 - \lambda^2 \cot^2\theta_0)^{1/2}. \tag{5}$$

If the matter at the point of emission is moving in a circular orbit in the equatorial plane of the black hole the ratio of frequency seen by a distant observer to frequency in the comoving frame at emission, the frequency shift factor g, depends on the angle the beam makes with the $+\phi$-direction, but is independent of the other direction cosines. In turn, the direction cosine with the $+\phi$-direction is independent of η.

Wilkins (1972) shows that a null geodesic can have at most one turning point. In the (η, λ)-plane the null geodesics with no turning points fill a compact region including the origin, and outside this region all the null geodesics have one turning point. The 'circular' null geodesic trajectories which stay forever at a constant value of r are on the boundary of the no-turning-point region and are obviously all unstable. A parametric pair of equations for the boundary is

$$\lambda = \frac{-r^3 + 3Mr^2 - a^2(r+M)}{a(r-M)}, \tag{6}$$

$$\eta^2 = \frac{r^3}{a^2(r-M)^2}[4a^2M - r(r-3M)^2], \tag{7}$$

and circular null trajectories are possible in the range of r for which $\eta^2 > 0$ (Bardeen, 1973). A black hole in front of a bright source of much larger angular size, like a black hole in a binary system passing in front of a large bright star, looks like a black disk, with a radius of about $5M$, whose boundary in the (α, β)-plane corresponds to the locus of circular photon orbits in the (η, λ)-plane. Unfortunately, there seems to be no hope of observing this effect.

The appearance of a point source of radiation which radiates isotropically in its own rest frame as it moves in a circular geodesic orbit in the equatorial plane of an extreme $(a = M)$ Kerr black hole has been calculated by Cunningham and Bardeen (1973) for several orbital radii and for observers at several polar angles. There are an infinite number of images visible at any one time; for each image the energy flux is the product of its surface brightness, proportional to $[g(\lambda)]^4$, and its apparent angular size. Images can be classified by the number of times the beam of radiation crosses the equatorial plane between the source and observer. Images corresponding to more than one or two crossings are negligibly faint, but if the source orbits at $r \lesssim 3M$ the one-crossing image is brighter on the average than the zero-crossing image. The time variations in the energy flux from the image seem to be primarily due to variations in surface brightness as $g(\lambda)$ varies; the variations are quite dramatic if the source is at $r \lesssim 20M$ and the observer is within about 30° of the equatorial plane.

These results would apply directly to something like a pulsar orbiting a massive black hole in the center of the Galaxy. An ordinary star wouldn't do, both because it would be torn apart by tidal forces unless the black hole were very massive ($M \gtrsim 10^8$ $M_\odot$) and because its optical radiation would be absorbed by intervening dust. Unfortunately, pulsars are rare objects in the Galaxy, and while perhaps not completely inconceivable, it seems highly unlikely one would find itself sufficiently close to the hypothetical black hole.

If applied to accretion disks, the results do seem to confirm the possibility of large fluctuations in brightness from hot spots orbiting in the inner part of the disk. The beams of radiation which cross the equatorial plane will mostly be reabsorbed or scattered if a/M is fairly close to one, which suggests that a good fraction of the radiation emitted inside $r \simeq 3M$ will not escape directly to infinity. The whole energy balance of the innermost part of the disk will be substantially modified by the effect of the strong gravitational field on the propagation of the radiation. Also, the flaring of the disk far from the black hole (Shakura and Sunyaev, 1973) will be considerably enhanced over Newtonian estimates by the tendency of the gravitational field to focus radiation emitted by the inner part of the disk toward the equatorial plane. Cunningham (1973) has been exploring some of these possibilities.

If it happens that the orbital angular momentum of the binary system is not in the same direction as the spin angular momentum of the black hole, so the orbital plane of the accreting matter far from the black hole is tilted with respect to the equatorial plane of the black hole, the Lense-Thirring effect can have a big influence on the structure of the accretion disk even for $r \gg M$. If we take the polar axis to be along the axis of symmetry of the black hole, the Lense-Thirring effect causes the plane of a circular geodesic orbit at $r \gg M$ to precess with an angular velocity

$$d\phi/dt \simeq 2GJc^{-2}r^{-3} \tag{8}$$

about the axis of symmetry of the black hole. In the case of an accretion disk the precession of the plane of the orbit of a fluid element is accompanied by a slow radial drift inward at a rate, in the middle region of Novikov and Thorne (1973),

$$-v^r \sim r^{-2/5}. \tag{9}$$

Assuming a stationary flow, we get an unique precession angle at each radius,

$$\phi_p(r) = 2GJc^{-2} \int_\infty^r (-v^r)^{-1} r^{-3} \, dr \sim r^{-8/5}. \tag{10}$$

As r decreases the accretion disk gets more and more twisted, in the manner shown in Figure 3. The disk is sliced by the plane of the two angular momentum vectors. The scale of radius in Figure 3 depends on the coefficient of proportionality in Equation (9). For black holes in binaries $\varrho = 1$ might correspond to about $1000 M$. Typically, then, the twisting of the accretion disk becomes appreciable at a few hundred times the radius of the black hole, where the gravitational field is still Newtonian to a good approximation.

The picture drawn in Figure 3 neglects any torques due to viscosity. Crude estimates made by myself and J. Pettersen indicate that one component of the viscous torque tends to oppose the Lense-Thirring precession and another component tends to align the angular momentum of the orbiting material with that of the black hole. We expect that once the total precession angle has become moderately large the accretion disk will have relaxed into the equatorial plane of the black hole.

If the misalignment of the black hole with the orbital plane of the binary is fairly substantial, a sizable fraction of the energy released near the black hole may be redeposited in the disk in the region where the transition to the equatorial plane of the black hole occurs.

Whether or not misalignment can occur in binary systems like Cygnus X-1 is a complicated astrophysical question. Even if it is rather unlikely, the possibility of isolating an important relativistic effect warrants further examination.

3. Energy Extraction From Black Holes

Penrose (1969) was the first to suggest that rotating black holes can lose energy even though they accrete only matter and radiation with positive energy as measured by any local observer. Inside the ergosphere there exist physically realizable null and timelike geodesic trajectories with negative energy parameter E. When such a test particle falls into the black hole, the gravitational mass of the black hole decreases. The result is that the black hole has given up some 'rotational energy' to whatever injected the test particle into the negative energy orbit.

Wheeler (1971) has suggested that if tidal forces disrupt a star in the ergosphere part of the material might find itself in such negative energy orbits, and the rest of the material might escape with more energy than the total rest mass energy of the original star. Mashoon (1972) has studied tidal disruption in the Kerr geometry without coming to any firm conclusions. However, Press (Bardeen et al., 1972)

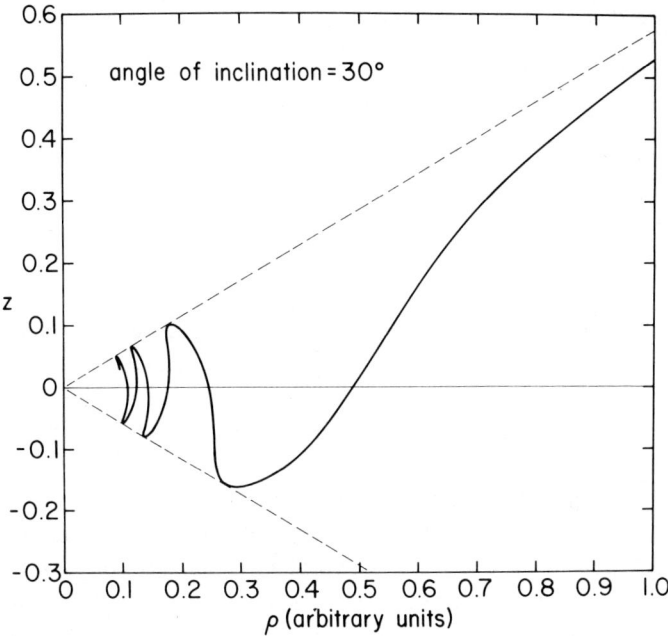

Fig. 3. A cross-section of an accretion disk which far from the black hole is tilted at an angle of 30° with respect to the equatorial plane of the black hole. The units of the cylindrical coordinates ϱ and z are scaled to show the beginning of the twisting of the accretion disk by the Lense-Thirring dragging of inertial frames to best advantage. The tendency of the viscous forces to relax the accretion disk into the equatorial plane of the black hole is not taken into account; they are only used to calculate the radial velocity of the matter as in Novikov and Thorne (1973).

has shown that any negative energy orbit has a velocity of at least $0.5c$ relative to any orbit which is accessible at $r \gg M$. It seems highly unlikely that tidal disruption can result in relative velocities much greater than the speed of sound in the original star, which rules out everything except neutron stars falling into small black holes.

Misner (1972) pointed out that waves can also extract energy from rotating black holes. Consider an electromagnetic or gravitational wave mode wich has frequency ω and axial angular eigenvalue m, so the wave amplitude is

$$\psi \sim W(r, \theta) \, e^{im\phi} \, e^{-i\omega t}. \tag{11}$$

The relationship between a change in energy and a change in angular momentum of the black hole is

$$dM = \omega m^{-1} \, dJ. \tag{12}$$

On the other hand, dM, dJ, and the change in area dA are related by (Hartle and Hawking, 1972)

$$dM = (8\pi)^{-1} \kappa \, dA + \Omega_H \, dJ, \tag{13}$$

where κ is the 'surface gravity' of the black hole and Ω_H is the angular velocity of the horizon (see Bardeen et al., 1973). Combining Equations (12) and (13) gives

$$dM = (8\pi)^{-1} \omega (\omega - m\Omega_H)^{-1} \kappa \, dA. \tag{14}$$

Since the area of the black hole can only increase (Hawking, 1971), if

$$0 < \omega < m\Omega_H, \tag{15}$$

the black hole mass decreases. If such a mode scatters off a rotating black hole the outgoing energy flux is greater than the ingoing energy flux, hence superradiant scattering. Numerical calculations by Teukolsky and Press (Teukolsky, 1973) indicate that the superradiance is less than 4% for electromagnetic waves, though Starobinsky and Churilov (1973) have shown analytically it can be as much as 137% for gravitational waves. No very plausible mechanism has been proposed for superradiance as such manifesting itself in a noticeable way in astrophysics (but see Press and Teukolsky (1972) on the 'black hole bomb').

Perhaps the astrophysically most interesting possibility is the existence of 'floating orbits' (Press and Teukolsky, 1972). From Equation (14) an orbiting test particle that has an angular velocity $\Omega < \Omega_H$ will extract energy from the black hole. If the energy extracted from the black hole exceeds the energy radiated to infinity, instead of spiralling into the black hole the particle will spiral outward until there is a balance; the particle then 'floats' at constant radius as it acts as a catalyst for extracting energy from the black hole. The calculations to check this possibility for electromagnetic or gravitational waves have not yet been done, though the mathematical techniques are now at hand, thanks to Teukolsky (1972).

4. Conclusion

My discussion has focussed on what relativistic effects may have observable consequences for models of accretion disks around black holes in binary systems, since at the present time there are the only systems in which there seems to be positive evidence for the existence of black holes. Black holes with accretion disks may also be the source of some of the non-thermal activity in galactic nuclei, from the modest level in the Galaxy to the most powerful quasars. Particularly for quasars, and probably in some X-ray binaries, the accretion rate may approach or even exceed the Eddington limit. The thin disk approximation, in which the fluid elements move in nearly geodesic trajectories in the equatorial plane, breaks down near the black hole when the luminosity approaches the Eddington limit. What then happens has not been explored in any detail, though Shakura and Sunyaev (1973) have given a qualitative picture. Particularly if the black hole is rapidly rotating any quantitative analysis will have to take into account strong relativistic effects.

Black holes also provide a number of mechanisms for generating gravitational waves, some of which are certainly of astrophysical interest. Such mechanisms are probably the only hope for explaining Weber's graviational wave experiments, if this is possible at all with conventional general relativity.

Through my whole discussion I have implicitly assumed that Kerr black holes are stable. Numerical calculations of Press and Teukolsky (1973) indicate rather strongly that the Kerr metric is stable to gravitational and electromagnetic perturbations, but a completely airtight proof of stability is not yet at hand.

What I hope my talk illustrates is that general relativity is more than a background in which to do astrophysics; rather, the relativistic aspects of black holes can generate qualitatively new observational effects. Much more theoretical work is necessary to develop the detailed models which will hopefully make possible a new type of confrontation of general relativity theory with observation.

References

Bardeen, J. M.: 1970, *Nature* **226**, 64.
Bardeen, J. M.: 1973, in C. DeWitt and B. DeWitt (eds.), *Black Holes*, Gordon and Breach, New York, London, Paris, p. 229.
Bardeen, J. M., Carter, B., and Hawking, S. W.: 1973, *Commun. Math. Phys.* **31**, 161.
Bardeen, J. M., Press, W. H., and Teukolsky, S. A.: 1972, *Astrophys. J.* **178**, 347.
Bardeen, J. M. and Wagoner, R. V.: 1971, *Astrophys. J.* **167**, 359.
Carter, B.: 1968, *Phys. Rev.* **174**, 1558.
Carter, B.: 1971, *Phys. Rev. Letters* **26**, 331.
Cunningham, C. T. and Bardeen, J. M.: 1973, *Astrophys. J.* **183**, 237.
Cunningham, C. T.: 1973, Ph.D. Thesis, University of Washington.
Hartle, J. B. and Hawking, S. W.: 1972, *Commun. Math. Phys.* **26**, 87.
Hawking, S. W.: 1971, *Phys. Rev. Letters* **26**, 1344.
Hawking, S. W. and Ellis, C. F. R.: 1973, *The Large Scale Structure of Space-Time*, Cambridge University Press, Cambridge.
Israel, W.: 1967, *Phys. Rev.* **164**, 1776.
Kerr, R. P.: 1963, *Phys. Rev. Letters* **11**, 238.
Lynden-Bell, D.: 1969, *Nature* **223**, 690.
Lynden-Bell, D. and Rees, M. J.: 1971, *Monthly Notices Roy. Astron. Soc.* **152**, 461.
Mashoon, B.: 1972, Ph.D. Thesis, Princeton University.
Misner, C. W.: 1972, *Phys. Rev. Letters* **28**, 994.
Newman, E. T., Couch, E., Chinnapared, R., Exton, A., Prakash, A., and Torrence, R.: 1965, *J. Math. Phys.* **6**, 918.
Novikov, I. D. and Thorne, K. S.: 1973, in C. DeWitt and B. DeWitt (ed.), *Black Holes*, Gordon and Breach, New York, London, Paris.
Peebles, P. J. E.: 1972, *Gen. Rel. Grav.* **3**, 63.
Press, W. H. and Teukolsky, S. A.: 1972, *Nature* **238**, 211.
Press, W. H. and Teukolsky, S. A.: 1973, *Astrophys J.* **185**, 649.
Pringle, J. E. and Rees, M. J.: 1972, *Astron. Astrophys.* **21**, 1.
Ruffini, R. and Wheeler, J. A.: 1970, in A. F. Moore and V. Hardy (ed), *The significance of Space Research for Fundamental Physics*, European Space Research Organization, Paris.
Shakura, N. I. and Sunyaev, R. A.: 1973, *Astron. Astrophys.* **24**, 337.
Shapiro, S. L.: 1973, *Astrophys. J.* **180**, 531.
Shvartsman, V. F.: 1971, *Soviet Astron. AJ* **15**, 277.
Starobinski, A. A. and Churilov, S. M.: 1973, *Zh. Eksperim. Teor. Fiz.* **65**, 3.
Sunyaev, R. A.: 1972, *Astron. Zh.* **49**, 1153.
Teukolsky, S. A.: 1972, *Phys. Rev. Letters* **29**, 1114.

Teukolsky, S. A.: 1973, Ph.D. Thesis, California Institute of Technology.
Thorne, K. S.: 1974, *Astrophys. J.*, to be published.
Wheeler, J. A.: 1971, in D. J. K. O,Connell (ed.), *Nuclei of Galaxies*, North-Holland Publishing Co., Amsterdam, London.
Wilkins, D. C.: 1972, *Phys. Rev.* **D5**, 814.
Zel'dovich, Ya. B. and Novikov, I. D.: 1971, *Relativistic Astrophysics*, Vol. I, University of Chicago Press, Chicago.

PART III

ACCRETION OF MATTER AND X-RAY SOURCES

BINARY X-RAY SOURCES

RICCARDO GIACCONI

Center for Astrophysics,
Harvard College Observatory/Smithsonian Astrophysical Observatory,
Cambridge, Mass. 02138, U.S.A.

Abstract. The observational data concerning binary X-ray sources is reviewed, with emphasis placed on Her X-1, Cen X-3 and Cyg X-1. In particular, the evidence for the identification of Cyg X-1 as a black hole is discussed.

1. Introduction

With the advent of orbiting X-ray observatories, such as UHURU, X-ray astronomy has entered an era of exciting new astronomical discoveries. In particular, with respect to galactic X-ray astronomy, the discovery of pulsating X-ray sources in binary systems by UHURU, has given us a powerful new tool with which to study the physical processes occurring in stars near the end point of their stellar evolution.

Long before many of the results, I will be discussing, had been obtained, Zel'dovich and Novikov (1964) had suggested that condensed stars could be found as X-ray sources accreting matter from binary companions.

The first indication that some X-ray sources might be associated with collapsed stars was provided by the first optical studies of compact galactic sources. The spectrum of Sco X-1 (Sandage et al., 1966) was found to be similar to that of an old nova. It is well known that such objects are close mass exchange binaries containing a collapsed (probably white dwarf) companion (Kraft, 1964). Shortly thereafter, Burbidge et al. (1967), reported evidence for the binary nature of Cyg X-2. These observations suggested an attractive model for X-ray sources: matter lost from the primary star in a close binary would accrete onto a compact object, releasing ~ 1–10 keV of gravitational potential energy per particle in the case of a white dwarf, ~ 100 MeV per particle in the case of neutron stars, and black holes. The thermalization of this energy, either by shock waves or by viscous heating in a disk, would produce a hot gas which would emit X-rays. In this way, X-ray luminosities in the range 10^{36}–10^{38} erg s^{-1} could be explained by modest accretion rates in the range 10^{-8}–10^{-10} $M_\odot$ yr^{-1}. This idea was pursued by Cameron and Mock, Shklovsky, Prendergast and Burbidge in papers published in 1967–1968. However, further observations of Sco X-1 and Cyg X-2 failed to show definite evidence of binary motion. Thus, for a time the possibility that compact X-ray sources could be binaries decreased in popularity. A number of other models, based for the most part on analogy with pulsars, were proposed, and the picture became somewhat muddled. Then in 1971 a dramatic breakthrough in our understanding of compact X-ray sources was achieved as a result of observations by the X-ray satellite, UHURU. Two periodically pulsating X-ray sources, Cen X-3 and Her X-1, were discovered. The observation of eclipses and a Doppler variation in their period conclusively established their binary nature on the basis of X-ray data alone. Today, as a result of combined radio, optical and X-ray observations, we can make a fairly convincing case that all compact X-ray sources not associated with supernova remnants are associated with mass transfer binaries containing a collapsed star. Perhaps, most significant of all, in the case of Cyg X-1, X-ray astronomy has furnished the strongest evidence yet for the existance of a new class of objects, black holes.

The work by Kippenhahn and his associates, and by Paczyński, has shed considerable light on the evolutionary tracks that may lead to the formation of the type of binary systems which the X-ray observations seem to require (Paczyński, 1971). It was shown that in the evolution of short period binary systems, there exists the possibility of obtaining systems in which a collapsed star is accompanied by a massive companion which overflows it s lobe of the zero velocity surface. Detailed computations were carried out by van den Heuvel and Heise for Cen X-3 (van den Heuvel, 1972). They follow the evolution of a system containing stars of mass 16 $M_\odot$ and 3 $M_\odot$ and $\tau = 3$ days. Using the evolutionary track of Paczyński, they find a descendant system containing 15 $M_\odot$ and 4 $M_\odot$ with a period of 1.53 kays. Assuming a supernova explosion ejecting 3.5 $M_\odot$, and assuming that the system remains bound, they are left with a 15 $M_\perp$ star accompanied by a 0.5 $M_\perp$ neutron star with a period of about 2 days. It should be pointed out that several assumptions on the conservation of angular momentum and total mass for the system, as well as on the details of the explosion are made in the computation. Colgate has pointed out the difficulties of retaining a bound system through the supernova event and has questioned whether neutron stars can be found in binary systems at all (Colgate, 1968). It is not my intention to deal in detail with evolutionary questions which will be more adequately covered by other authors in Session 66, except to point out that the existence of an X-ray emitting phase appears to be a required stage in the evolution of massive stars in short period binaries, rather than an oddity.

Apart from the insight that the discovery of such systems gives us on the evolution of stars, the finding of compact objects in binaries presents us for the first time with the opportunity to investigate their properties in detail. Thus, the possibility exists in the case of Her X-1 and Cen X-3 that a precise determination of mass could be obtained, as in the case of double line spectroscopic binaries. In this case, the velocity of one component is derived by X-ray measurements and of the other by conventional spectroscopic techniques in the optical. If these objects are indeed neutron stars, a statistical analysis on several such systems will give us an important indication of the possible upper limits on their mass. Also, the detailed study of the changes in orbital period and pulsation periods which are observed in X-rays will greatly improve our understanding of matter loss processes from the system, changes in moments of inertia of neutron stars and the nature of their physical state (Lamb, 1972; Baker, 1973). The detailed analysis of X-ray absorption features, will permit us to understand the gas dynamics in the system. Finally, the discovery and subsequent study of black holes is of great significance for general relativity, a point vigorously made by Wheeler and Ruffini (1971).

In this survey of X-ray binaries, I will not attempt to expand on the points above. I will simply endeavor to bring to your attention the wealth of information which is rapidly being uncovered by X-ray observations.

2. Her X-1

The best studied of the binary X-ray sources is Her X-1. As we will see, many of the observational phenomena for Her X-1 can be clearly explained in terms of a rotating neutron star orbiting in a binary system, although some of the complex features of this system are not yet fully understood. The view that Her X-1 is a neutron star is by no means generally accepted. Models based on differentially rotating degenerate dwarfs have recently been suggested; and the proponents of vibrating degenerate dwarf have not yet conceded. Without going into the merits of the models, I have adopted the neutron star point of view as a means of more conveneintly describing the data. If, indeed, degenerate dwarfs with masses of 1.4 $M_\odot$ can rotate or pulsate at 1.24 s, then the alternate models are at least plausible although no detaled computation to explain the details of the observations has, to my knowledge, yet been carried out.

If we start from the simplest X-ray observations, we find that Her X-1 shows periodic occultations with a 1.7 day period. Figure 1 shows the 2–6 keV X-ray intensity, as observed by UHURU, varying from a high of 100 counts s^{-1} to a level below the limit of UHURU's detectability of a few counts s^{-1}. The transitions between the high intensity state and eclipse take a time less than 12 min. Data for 3 adjacent occultation cycles in July 1972 are shown in the figure, which also shows that the X-ray eclipse lasts for 0.24 days. Figure 2 from Forman *et al.* (1972), shows the optical behavior of Hz Herculis in the summer of 1972 and on plates from the 1940's. The 1.7 day optical variations in phase with the schematically represented X-ray eclipses of Her X-1 make certain the identification of the star with the X-ray source. The optical variations are interpreted in terms of a binary in which the central star has a hot side facing the X-ray source and heated by the X-ray emission with temperatures of order 10000 K for the hot side and 6000–7000 K for the cooler side. Recent papers, such as those of Wilson, Joss *et al.*, and Rucinski, have considered the various detailed models which are required to reproduce the shape of the observed optical variations.

The first X-ray observations of Her X-1 also showed another regular feature. The source pulses periodically with a 1.24 s period. This behavior is illustrated in Figure 3, where the lighter histogram shows 30 s of 2–6 keV counting rate data accumulated in 0.096 s bins, the highest time resolution available with UHURU. The heavier curve is a minimum X^2 fit to the pulsations of a sine function plus two harmonics.

Since Her X-1 undergoes regular eclipses, it should not be surprising that its 1.24 s pulsing shows a regular Doppler pattern. As the X-ray source orbits its binary companion, the pulses appear closer together (shorter period) as the source comes out of elipse and approaches us and the pulses appear farther apart (longer period) as the source heads towards eclipse and moves away from us. This behavior allows us to measure directly and precisely very many of the parameters of the Her X-1 system.

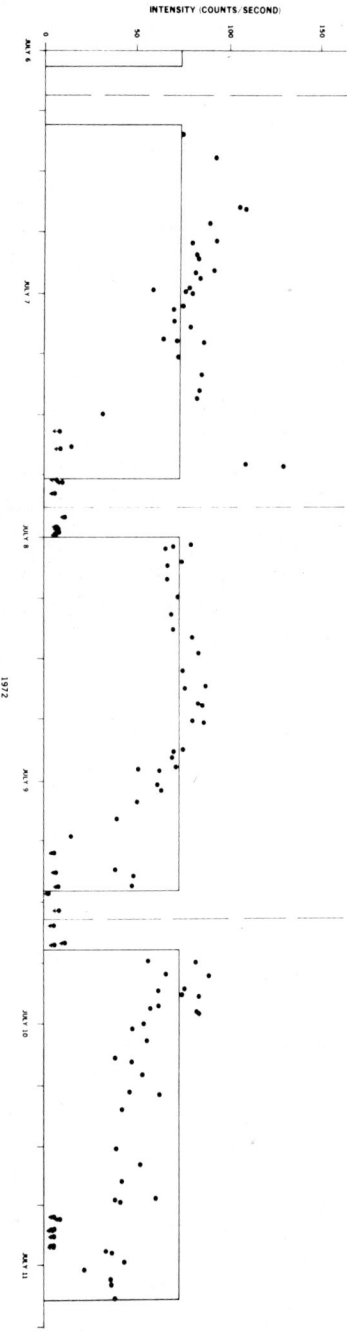

Fig. 1.

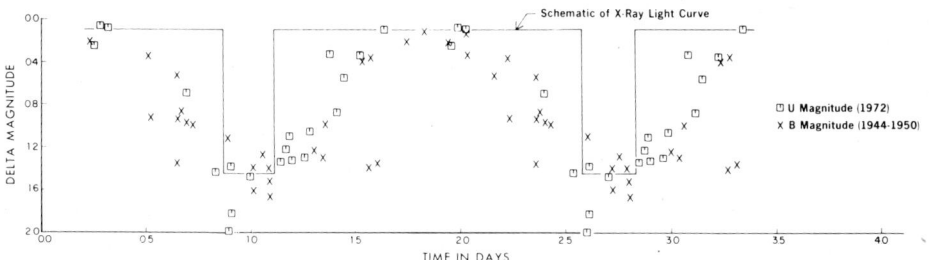

Fig. 2. Hz Herculis optical observations folded 1.70015 days.

The Doppler curve is indicated in Figure 4, where in the bottom half, we have replotted the intensity data from the first Hercules figure and in the top half, we have shown the time difference between the time of arrival of a pulse and the time predicted for a constant period plotted versus time. As can be seen from the figure, the arrival time of the pulse is delayed 13.2 s at the center of the occultation and is 13.2 s early at the center of the high state.

The delay time of 13.2 s is a direct measure in light seconds of the radius of the orbit of the X-ray star about the center of mass of the binary system as projected into the observing plane. Most importantly, the presence of the sinusoidal Doppler curve confirms the picture of the X-ray source orbiting in a binary system as first suggested by the regular eclipse.

Table I contains a summary of the observed parameters of the Hercules X-ray system – the average pulsation period τ, the half-amplitude of the period variation

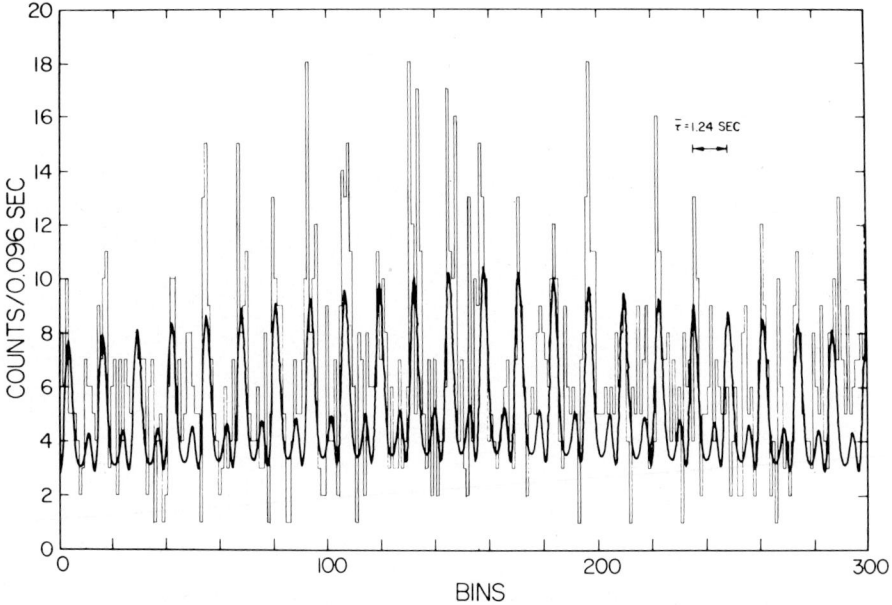

Fig. 3. Source in Hercules (2 U 1702+35). November 6, 1971.

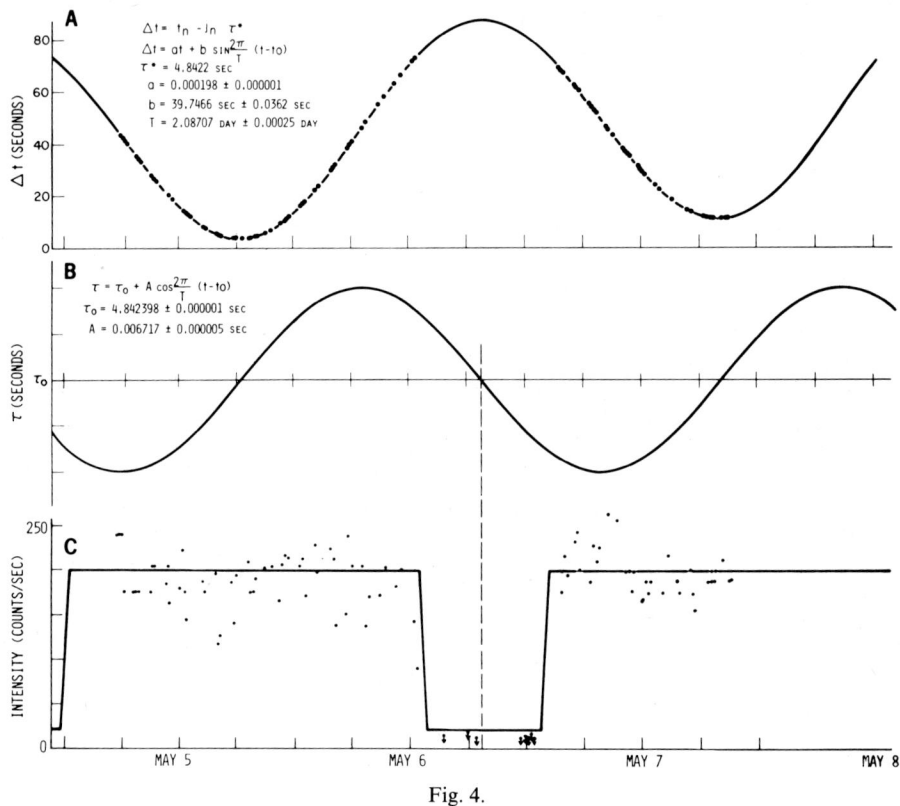

Fig. 4.

TABLE I

PARAMETERS OF HERCULES X-1

Observed Parameters

τ (average heliocentric pulsation period in seconds)	1.2378206 ± 0.0000001 (Jan. '72)
$\Delta \tau$ (half-amplitude of period variation in seconds)	0.0006979 ± 0.0000003
T (orbital period in days)	1.70017 ± 0.00001
ϕ_T (phase of eclipse: low center)	UT 1972 Jan. 13.0772 $\pm$ 0.0003 UT 1972 Jul. 6.1897 $\pm$ 0.0004

Derived Parameters

ϵ (eccentricity)	≤ 0.05
$v \sin i = (\Delta \tau / \tau) \, c$ (km s^{-1})	169.0 ± 0.1
$r \sin i = (T/2\pi) \, v \sin i$ (cm)	$(3.95 \pm 0.01) \times 10^{11}$
$M_2^3 \sin^3 i / (M_1 + M_2)^2 = (2\pi/T)^2 \, (r \sin i)^3 / G$ (grams)	$(1.69 \pm 0.01) \times 10^{33}$

$\Delta\tau$, the orbital period T, and the phase of the eclipse center ϕ_T. In addition, four paramaters are derived from the measured quantities under the model of a binary system in an orbit with an inclination angle i to observer ($i=90°$ corresponds to the observer being in the plane of the orbit of the binary system). The first derived parameter is the eccentricity of the orbit of the binary and from the good quality of the sine fit to the pulse arrival times is determined to be less than 0.05. This result justifies the assumption of a circular orbit which can be used to derive the projected orbital $v\sin i$, the projected orbital radius about the center of mass $r\sin i$, and the mass function of the system $(M_2^3\sin^3 i)/(M_1+M_2)^2$ where M_1 is the mass of the X-ray star and M_2 the mass of the occulting star.

We should point out that the value of the period T given in the table is the heliocentric value determined in January 1972. We have now determined the pulsation period for 14 times from December 1971 through March 1973. These data are shown in Figure 5 with corrections for all significant motions applied. The figure shows the heliocentric period vs time. Significant changes are clearly occurring but the picture is quite complex. From January 1972 to August 1972 the period decreased by $5\frac{1}{2}$ μs, from September 1972 to October 1972, the period increased by about 3 μs, and then from October 1972 to March 1973, the period decreased by $3\frac{1}{2}$ μs.

The fact that the X-ray period shows a net decrease of 6 μs over $1\frac{1}{4}$ yr rules out

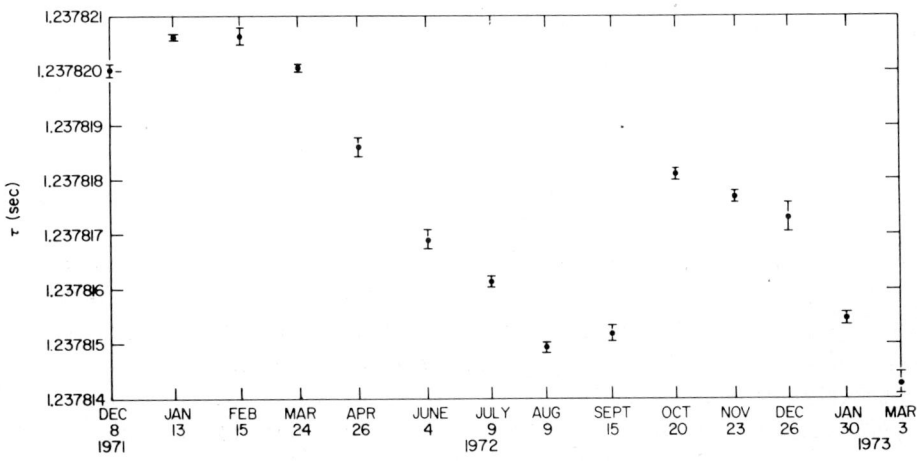

Fig. 5. Her X-1, pulsation period τ vs time.

models such as have been applied to radio pulsars and the Crab Nebula pulsar, in which a rotating neutron star slows down (period increases) at a rate approximately equal to that required to supply the energy emitted as radiation.

Since Her X-1 is a member of a binary system, we turn to accretion as the most likely energy source with the rotation of a neutron star still providing the clock mechanism. Although accretion models are extremely complicated, a few general statements may be appropriate. Material can be accreted radially if it has insufficient angular momentum for centrifugal forces to halt its free fall. However, for close binary systems accreting matter is likely to have so much angular momentum that it cannot fall directly onto the compact object. In this case, the matter will form a disk composed of material which gradually spirals inward and transfers its angular momentum to the compact star. The accretion problem has been studied by Zel'dovich and Novikov (1971), Prendergast and Burbidge (1968), Shvartsman (1971), Pringle and Rees (1972), Ostriker and Davidson (1972), Lamb et al. (1973), and Shakura and Sunyaev (1973). Most relevant to the X-ray results on the speeding up of Her X-1, both Pringle and Rees, and Lamb et al. predict that matter being accreted will transfer angular momentum to the star thereby speeding it up with the predicted changes being of the same order that we have observed for Her X-1. The models can explain the more complicated picture now presented on the change of the period with time on the basis of changes in the rates of matter transfer.

Distances determined from optical data imply an X-ray luminosity from 10^{36} to 10^{37} erg s^{-1} although this could decrease by a factor of 10 by accounting for the limited solid angle that the pulsed source fills. The accretion process is capable of producing at least 10^{38} erg s^{-1} and therefore is capable of producing the observed luminosity. As already described in the last table, the mass function of Her X-1 is 0.85 solar masses. One of the most important quantities to determine is the mass of the X-ray source itself which might yield the first direct measurement of the mass of a neutron star. The mass function derived from the Doppler shift of the X-ray pulsation period gives one equation involving M_1, M_2 and $\sin i$. There are several approaches that can be followed at this point. The simplest conceptually is to determine the velocity of the central star about the system center of mass from visible light and spectroscopic observations. Further the spectral typing of the central star allows its mass to be determined, in principle. These two additional results then determine with the X-ray data M_1, M_2 and i. Alternatively, the detailed shape of the optical eclipse can be used to determine the system parameters. These areas of effort are still the subject of much discussion and debate on the interpretation of the observed radial velocities, although preliminary results, such as those of Crampton and Hutchings, who obtain $1.4 \pm 0.4\,M_\odot$ for the X-ray star, agree with the numbers obtained below.

Another technique that can be used to estimate the masses involved uses the X-ray data alone. One equation relates the size of the Roche lobe of the primary to the separation of the stars and their masses. A second expression relates the radius of the occulting object, the separation of the binary components, the inclination angle i, and the phase angle of the occultation duration, a measured quantity. Using these equations requires us to relate the radius of the occulting region to the size of the Roche lobe and this is where some difficulty lies. Among others, Wilson has argued that the sharpness of the eclipse transition requires so large a density gradient that the radius of the occulting region cannot be larger than the Roche lobe. Ruffini, on the other hand, argues that the radius of the occulting region must be larger than the Roche lobes since the star is losing mass. Depending upon which inequality is correct, we can calculate a range of allowable masses for each value of i. If we assume, as is likely, that $R \approx R_L$, we obtain a value of M_1 and M_2 for each possible i, and some results of such a calculation are tabulated in Table II. For example, at an inclination of 85°, assuming the radius of the occulting region is equal to the size of the Roche lobe, determines a mass of 1.99 solar masses for the central star and 1.04 $M_\odot$ for the X-ray source.

TABLE II

HERCULES X-1 MASS ESTIMATES - X-RAY DATA ONLY

Assumes $R_{occulting} = R_{Roche\ lobe}$

Inclination Angle	Mass X-Ray Star	Mass Central Star
90°	1.19	2.09
85°	1.04	1.99
80°	0.74	1.78
75°	0.45	1.56
60°	0.09	1.47
45°	0.03	2.46
30°	0.04	6.8

The observations of Her X-1 show further structure in the X-ray intensity vs time. For 10 or 11 days, the source is intense and pulsing and can be seen following the 1.70 day occultation cycle. Then for 24 days the source is too weak to be observed.

Figure 6 shows the detailed X-ray intensity data, corrected for aspect, observed during the high states in January, March and July, 1972. Each dot represents a single spin of the satellite across the source. Two types of error bars are shown in the figure; the statistical error bar is determined from counting statistics and is the appropriate error bar to apply locally when considering variability. The larger error bar is dominated by the aspect correction and must be applied when comparing data taken on different days since the satellite is normally maneuvered once per day.

In addition to the fluctuations observed within a day, there are other features that stand out in this figure. First, we see that the source turns on rather abruptly. The intensity then increases rapidly to a maximum level and stays near this level for several days. The intensity then decreases rather smoothly for several days until the source is not detectable. Recently, the X-ray experiment on Copernicus has detected X-ray emission from Her X-1 during a time in which it was predicted to be in the off part of the cycle. We first established the 35 day cycle from observations of 5 cycles from November 1971 until March 1972; 4 of which the source was monitored with coverage on 67 out of 96 days occurring between high states. On all of these 67 days, as well as a number of additional scattered days of observation, the source was not detected

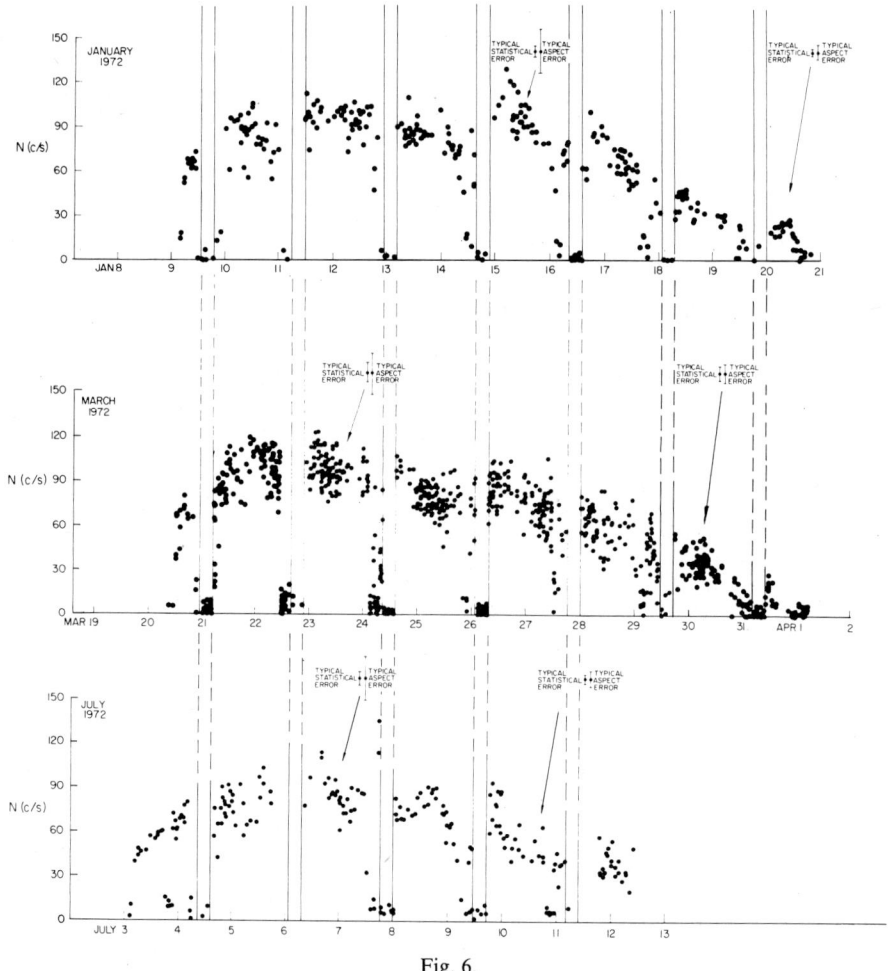

Fig. 6.

above background, thereby indicating the rarity of events, such as detected by Copernicus and possibly by the Livermore group in a May 1971 rocket flight.

In Figure 7 (top half) we have used the sharpness of the high state turn on to look at the time of occurrence of the 35-day cycle. We have plotted the time of the turn-on minus an integer *34.85. The first several points lie on a straight line whose slope indicates the 35.7 day periodicity first reported. Other points lie on lines with slopes corresponding to 34.85 day and 34.0 day periods, while a few points are scattered off the lines. While the turn-on data definitely do not show a strict periodicity, we should note that all of the points are within one eclipse cycle of the average 34.85 day cycle chosen. Thus, while there appears to be an underlying clock, it just does not seem to keep time precisely except on the average.

This may be understood, perhaps, by considering the bottom half of Figure 7 where we have plotted the turn-on times as a function of 1.7 day orbital phase. We should note that the range of phase indicated in this plot, as well as the error bars in the upper half of the figure, represent absolute limits on the uncertainty in turn-on times. We see that the turn-ons cluster at two phases: around phase 0.2 and 0.7. An earlier analysis suggested that all of the turn-ons could be consistent with two very well defined times, but as more data have been analyzed, we find that while the turn-ons do cluster near phases 0.2 and 0.7, there are some that definitely do not overlap.

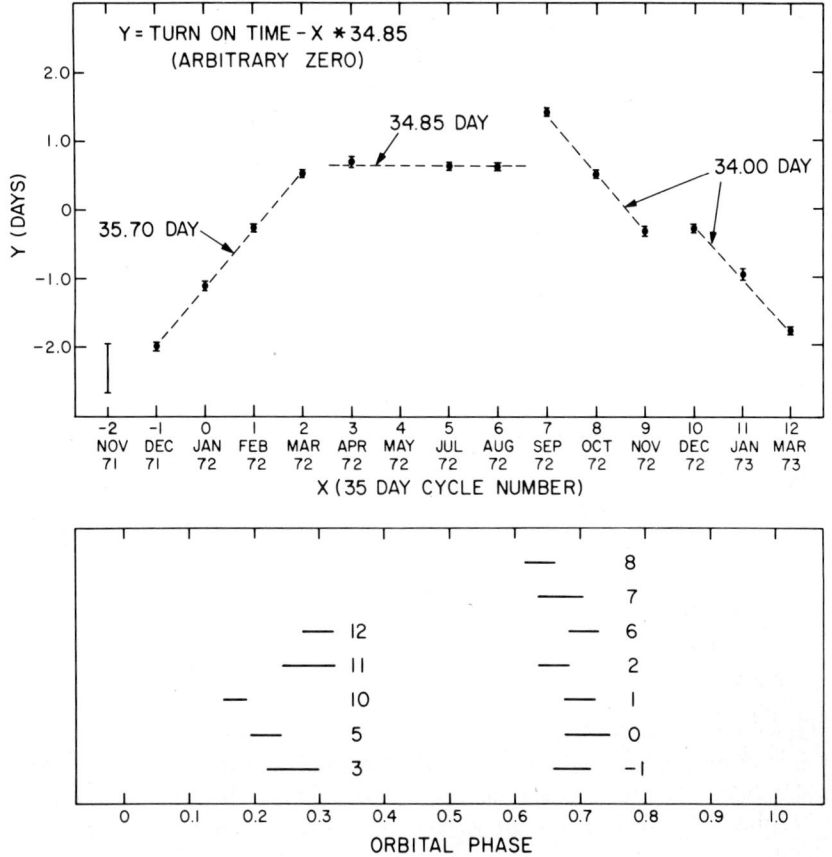

Fig. 7. Her X-1 35-day turn on.

The simplest model for the 35 day cycle is torque-free precession of an oblate spheroid neutron star first described by Brecher. If this is the case, then the 35 day cycle of Her X-1 is a reflection of conditions in the interior of the neutron star. For example, Pines, Pethick, Lamb and Shaham have shown that the model works only if the neutron star has a solid core. This model has been modified by Pines and co-workers to explain the sudden onset of the 'on' state by a triggering effect of the accreting material piling up at the Alfvén surface until the precession angle becomes such that acretion onto one or the other of the magnetic pole regions is no longer prevented. By assuming that the accumulated disk of matter acreted during the off state is thickest near the inner and outer Lagrange points, Pines and co-workers explain the turn-ons at phases near 0.25 and 0.75 as occurring at the locations at which the X-rays can most easily 'burn' through the disk and first be seen. Further detailed work is, of course, necessary to fully evaluate this model and to see if further details of X-ray absorption by streaming gas, such as the dips described by Giacconi *et al.* (1973a), and the complicated optical behavior can be explained by the model.

3. Cen X-3

We now turn our attention to Cen X-3. Figure 8 shows this source pulsing with a

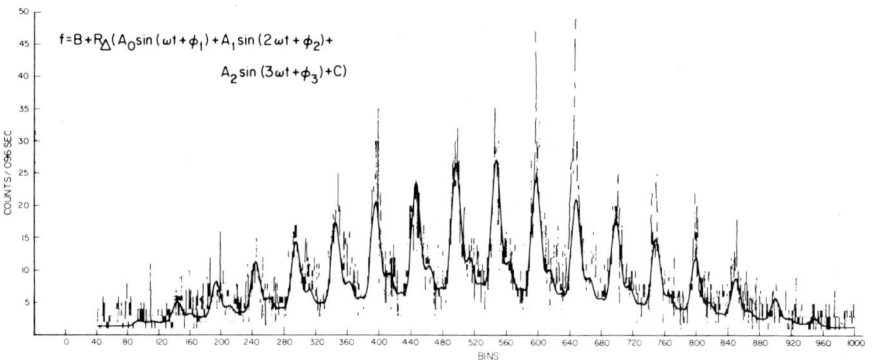

Fig. 8. Cen X-3 (2U 1119–60). May 7. 1971.

4.8 s period. The lighter histogram is again the 2–6 keV intensity observed with UHURU in 0.096 s bins and the heavier curve is a fit to the data. Figure 9 shows that, as for Her X-1, this source undergoes regular occultations and that the 4.8 s pulse period is Doppler shifted in phase with the eclipses. One difference from Her X-1 is that Cen X-3 does not exhibit a regular 35-day cycle, but as is shown in Figure 10, has times when it is intense and pulsing, and other times when it is seen weakly (if at all), and still other times when it alternates erratically between a high state and a low state. The data in the figure cover the $2\frac{1}{2}$ yr time over which UHURU has operated. There are no clear cut regularities in the highs and lows – no single period comes close to fitting all the observations. The data suggest that typical high states may last for around 4 months and typical low states for around 2 months; but exceptions are almost the rule.

In Figure 12a we have plotted several heliocentric 4.8 second periods observed for Cen X-3. The changes in period are so great that error bars are smaller than the dots representing the data points. The period decreased by 1.1 ms from January to May 1971, decreased by 0.2 ms from May to July, decreased by 0.25 ms from July to December 1971, decreased by 1.3 ms from December 1971 to September 1972 and increased by about 50 μs from September to October 1972.

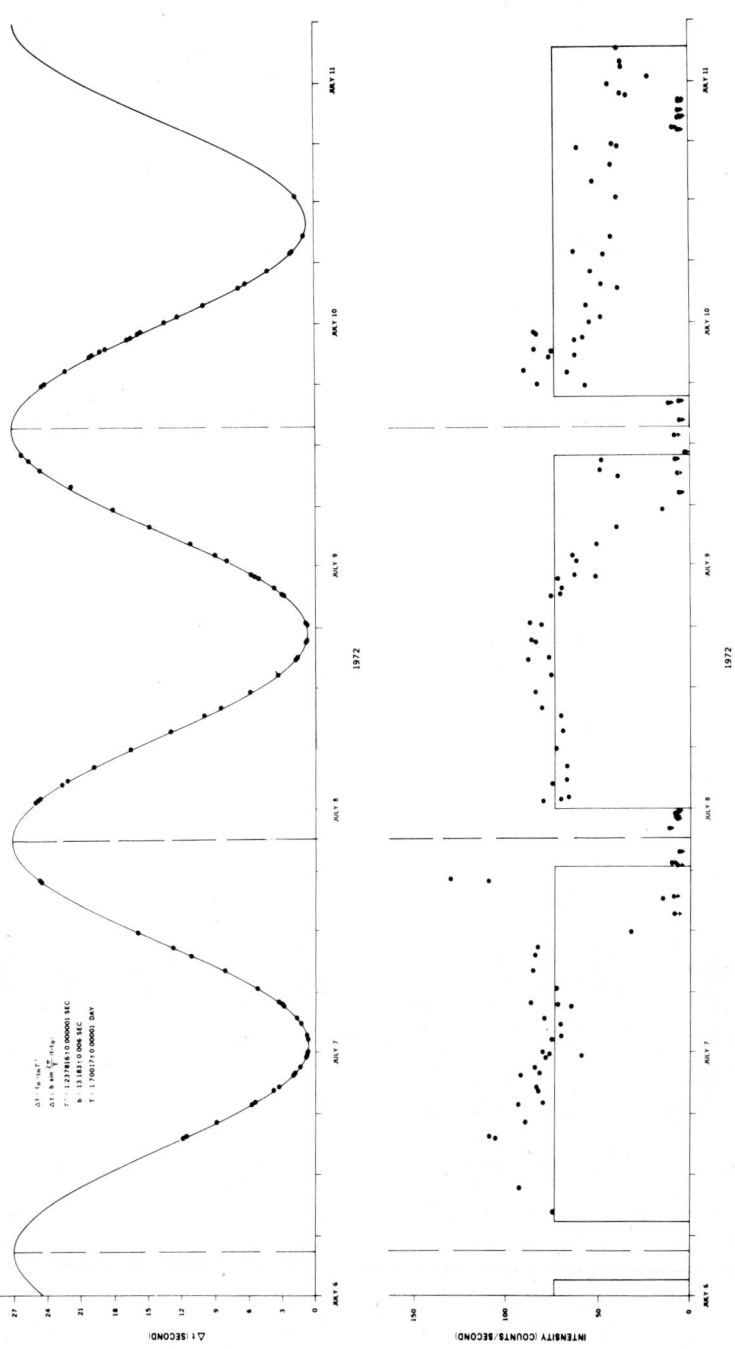

Fig. 9.

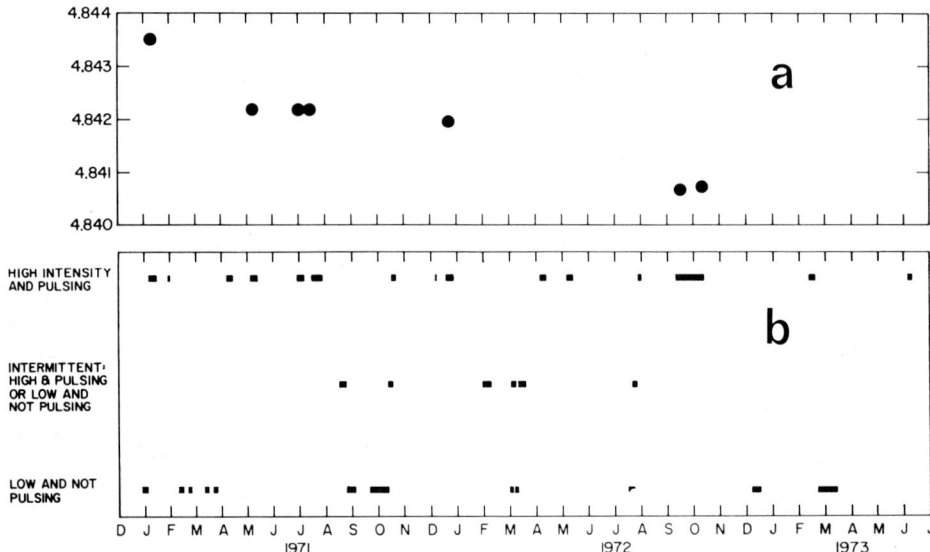

Fig. 10. (a) Cen X-3, pulsation period vs time. (b) Cen X-3, intensity vs time.

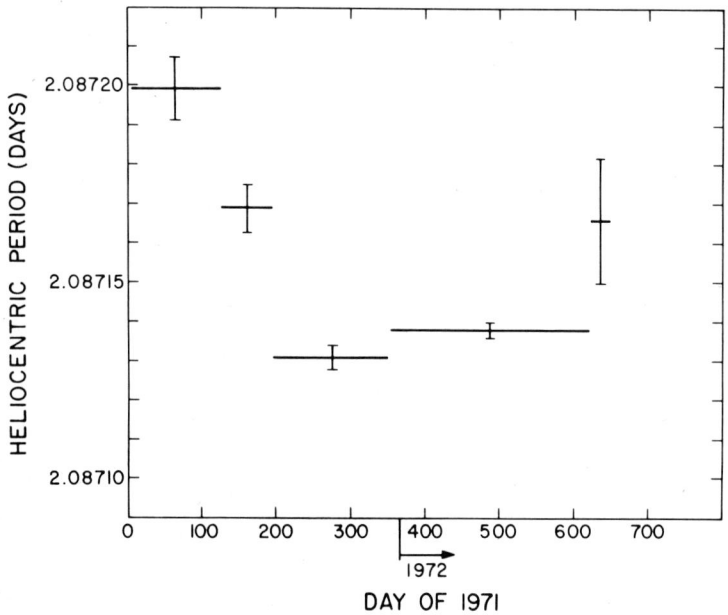

Fig. 11. Orbital period of Cen X-3 (averaged over indicated bxselines).

One additional piece of evidence in the Cen X-3 picture is shown in Figure 11, where we plot the two day orbital period of Cen X-3 vs time. The first few data point indicate that the orbital period is decreasing at a rate consistent with mass loss from the primary at a rate of $\sim 10^{-5}$ solar masses per year. However, additional data have shown that the orbital period is increasing with time. Liller has recently found a 14th mag. optical counterpart for Cen X-3, although the identification requires small changes in the presently observed X-ray period to fit the data on old plates. Changes that are required are of the order of the changes observed in the X-ray orbital period so at present the identification seems likely to be correct.

The dissimiliarities of their detailed behavior notwithstanding, Her X-1 and Cen X-3 seem to fit well in a model where accretion occurs on a magnetic, rotating neutron star of small mass 0.2 to 1.4 $M_\odot$. The observational data regarding Cyg X-1 seem, however, to require a different model.

4. Cyg X-1

The discovery of short X-ray pulsations from Cyg X-1 has been certainly one of the most significant achievements of the UHURU orbiting observatory with regard to galactic X-ray sources. This discovery stimulated a wealth of X-ray observations on periodic and non-periodic pulsating X-ray sources which resulted in the discovery of Her X-1, Cen X-3, and many of the binary sources we are presently studying. It also stimulated a concentrated effort in identifying the radio and optical counterparts of the object, whose detailed study has led to the conclusion that the Cyg X-1 system contains a black hole. Due to the importance of this finding and to the rapidly accumulating evidence strengthening this conclusion, I believe it is useful to review some of the experimental results and discuss more of the argument involved in reaching it.

Cyg X-1 has been observed in some of the earliest surveys in X-ray astronomy. In 1966, a survey of the Cygnus region (GR 67a) resulted in the first accurate location determinations for Cyg X-1, Cyg X-2 and Cyg X-3. As a result of this survey, the optical candidate for Cyg X-2 was found. No candidate objects could be found for Cyg X-3 or Cyg X-1. The energy spectra of the Cygnus sources were also measured. The spectrum of Cyg X-1 covering the range from 1 to 80 keV was measured by many observers from balloons and rockets. It appeared to have a flat power law spectral shape ($E^{-\alpha}$) with $\alpha=0.7$, thus rather similar to the one found in the Crab Nebula. It was considered puzzling that while the source was not too different in intensity and spectral form from the Crab Nebula source, no evidence of radio emission could be found with a limit of 1/500 of Crab radio flux. Since at the time we only knew of 2 types of X-ray sources, Sco X-1-like and supernovas, this was considered evidence for a different type of X-ray emitter. How truly different was not revealed to us until December, 1971, when the UHURU X-ray observatory detected the existence of X-ray pulsations from Cyg X-1 (Oda *et al.*, 1971; Figure 12).

Figure 13 contains data already reported in the literature (Schreier *et al.*, 1971) showing substantial variations in X-ray intensity on time scales from 100 ms to 10's of seconds. Some 80 s of data are shown here summed on 4 time scales from 100 ms up to 14 s. I should point out that similar X-ray variability also reported by

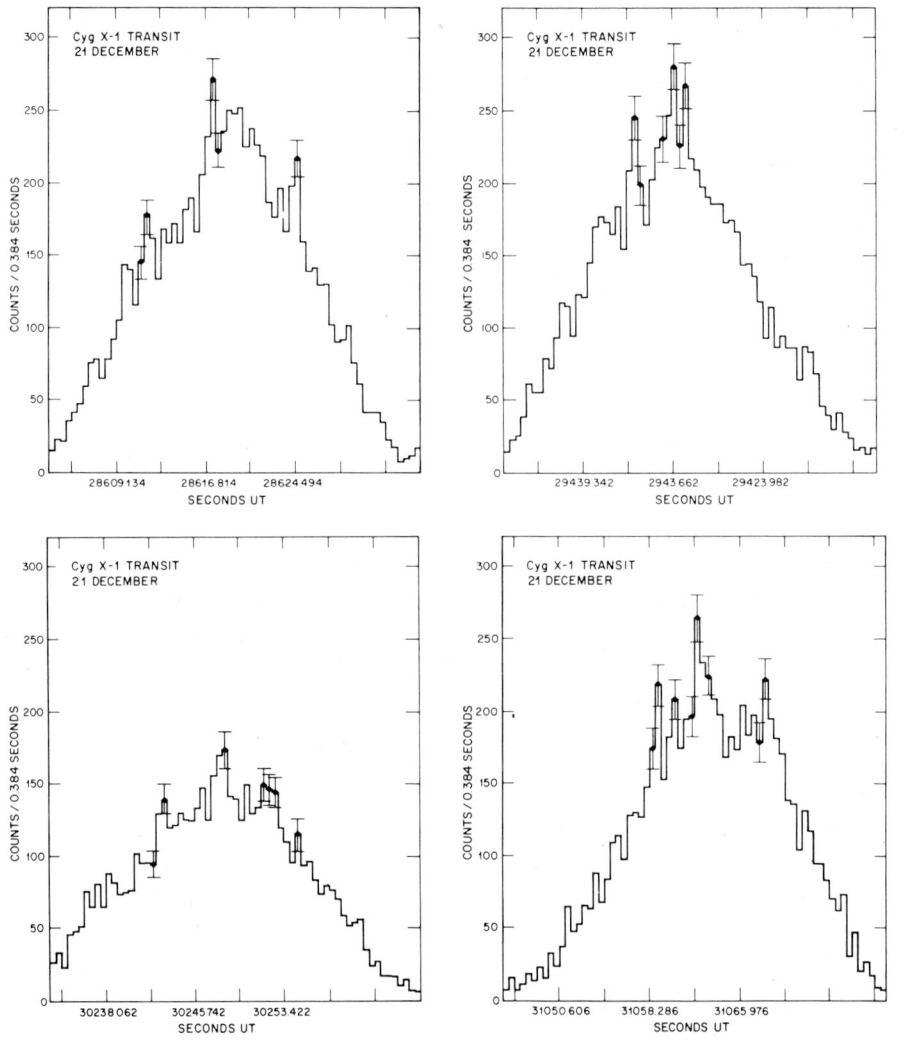

Fig. 12.

scientists at MIT (Rappaport *et al.*, 1971a), Goddard Space Flight Center (Holt *et al.*, 1971) and NRL (Shulman *et al.*, 1971) compels us to consider a source region of 10^9 cm or less.* Figure 14 shows the X-ray location obtained from an MIT rocket flight and from UHURU which led to the discovery of a radio source by Braes and Miley (1971) and by Hjellming and Wade (1971). It is this precise radio location that led to the optical identification by Webster and Murdin (1972) and by Bolton (1972) of Cyg X-1 with the 5.6 day spectroscopic binary system HDE 226868. The central object of this system is most likely a 9th mag. B0 supergiant and conservative mass estimates for the primary lead to a mass in excess of several $M_\odot$ for the unseen

* See note added in proof, p. 178.

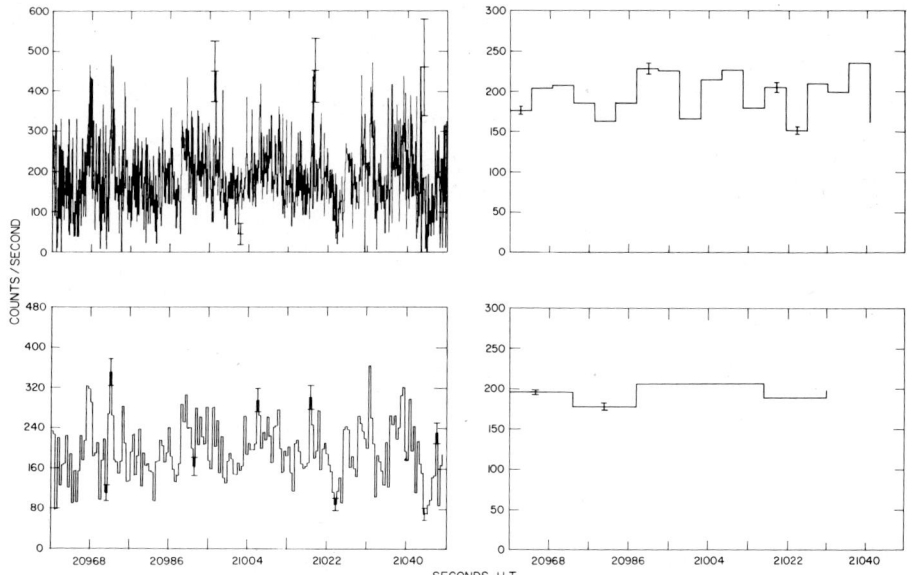

Fig. 13.

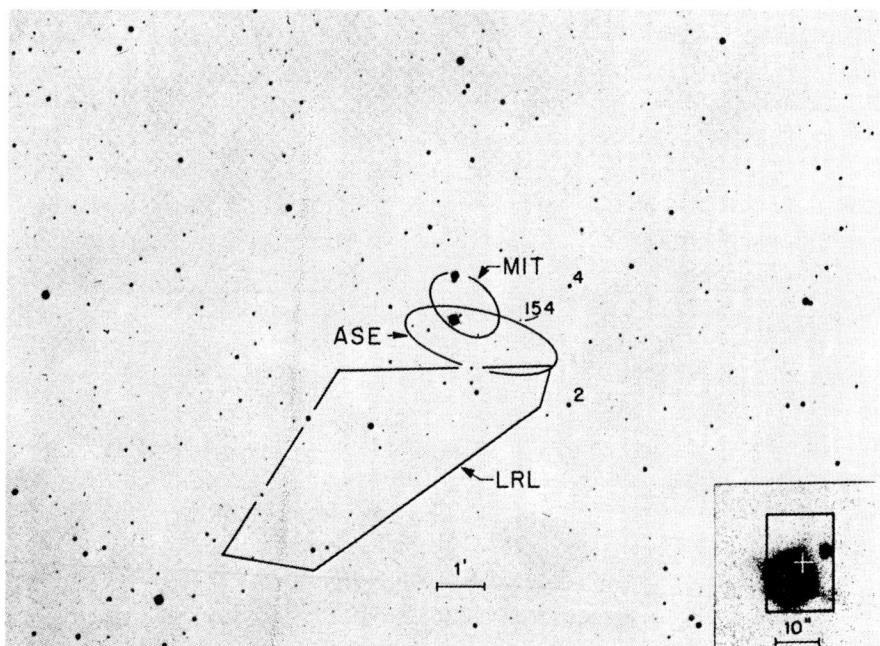

Fig. 14.

companion. If the companion is the compact X-ray source, then it could be a black hole.

The conclusion that Cyg X-1 consists of a binary system in which one of the stars is a black hole is based then upon three main points:
(1) HDE 226868 is the optical counterpart of Cyg X-1,
(2) The mass of the HDE 226868 $\geqslant 20\ M_\odot$, and
(3) The X-ray emitting object is compact.

I will consider the three points in order.

4.1. IDENTIFICATION

The identification is based on first order of the positional coincidence between X-ray and optical object (1′), Figure 14. No evidence, however, could be found in the Uhuru data at high X-ray energies (2–10 keV) of the binary nature of the system. Since the positional coincidence between radio and optical yields much greater accuracy (1″), we have attempted to establish a correlated behavior between radio and X-ray emission.

With the use of UHURU as an observatory, we have now analyzed 16 months of data on Cyg X-1 which are shown in Figure 15. We have plotted the 2–6 keV intensity vs day of 1970. The vertical lines for a given day show the range of variability observed on that day. For some days we have only the average intensity shown by a dash available in our analyzed results. We see that a remarkable transition occurred in March and April, 1971, with the source changing its average 2–6 keV intensity level by a factor of 4. We have also indicated in the figure the 6–10 keV and 10–20 keV X-ray intensities and see that the average level of 10–20 keV flux increased by a factor of 2. The figure also shows that at the same time the X-ray intensity changed, a weak radio source appeared at the Cyg X-1 location and was detected by the Westerbork and NRAO groups. Hjellming has recently reported analysis of additional radio data which shows the radio source first appeared sometime between March 22 and March 31, essentially the time during which the 2–6 keV X-ray intensity first headed downward. This correlated X-ray radio behavior is strong evidence, in addition to the positional data, that Cyg X-1 is in fact identified with the optical and radio object.

In Figure 16 we see the Cyg X-1 'average' spectrum before the March 1971 transition and after. The spectrum before the transition has a low energy excess which can be fit by either a power law with energy index of 4 or an exponential with a temperature of 11×10^6 K. The disappearance of this low energy component at the same time the radio source appeared, could be related to decrease in the plasma density which reduced the X-ray emission measure and also reduced the plasma frequency or the free-free absorption of the radio emission. Additional evidence for the identification has been reported by Sanford (1973) on the basis of Copernicus satellite data. He reports that a decrease of soft X-ray emission from Cyg X-1 was detected at Phase 0 of the 5.6 period of several occasions. Such low energy X-ray behavior would prove conclusively the identification.

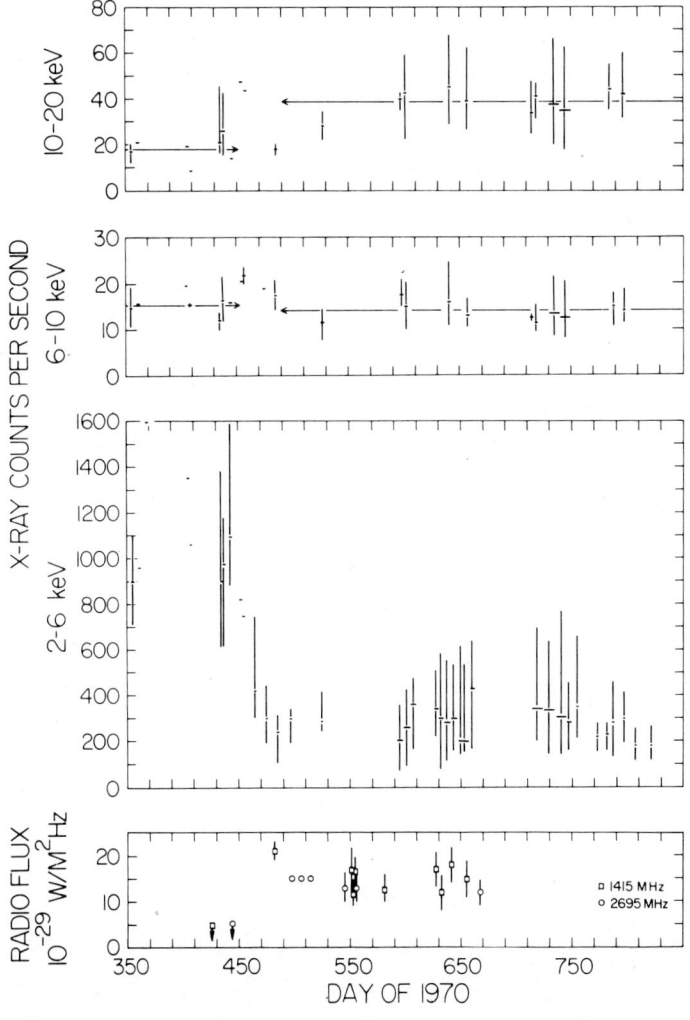

Fig. 15. Cyg. X-1.

4.2. Mass of HDE 226868

On the basis of the arguments given above, the HDE 226868 system is believed to be the optical counterpart of Cyg X-1. Bolton (1972) and Brucato and Kristian (1972) derived spectroscopically a mass function which leads to a minimum mass of about 5 $M_\odot$ for the X-ray emitting secondary star. This conclusion was based on the assumption that the primary star of HDE 226868 is a normal B0Iab supergiant of more than 20 $M_\odot$ according to the spectral characteristics given, for instance, by Walburn (1973). Paczyński (1972) and Trimble et al. (1973) criticized this assumption. Indeed a spectrum gives only information on the effective temperature and the surface gravity of a star. Trimble et al. (1973) gave a model of a low mass star which is in full agreement with the observed spectral type. Paczyński pointed out that the strong

X-ray flux from the secondary might alter the appearance of the observable spectrum of a star. Four recent papers have addressed these criticisms. One paper by Bolton (1973) used mass function as determined from the absorption line velocities, the HE II 4686 emission line velocities and the distance derived from interstellar red-

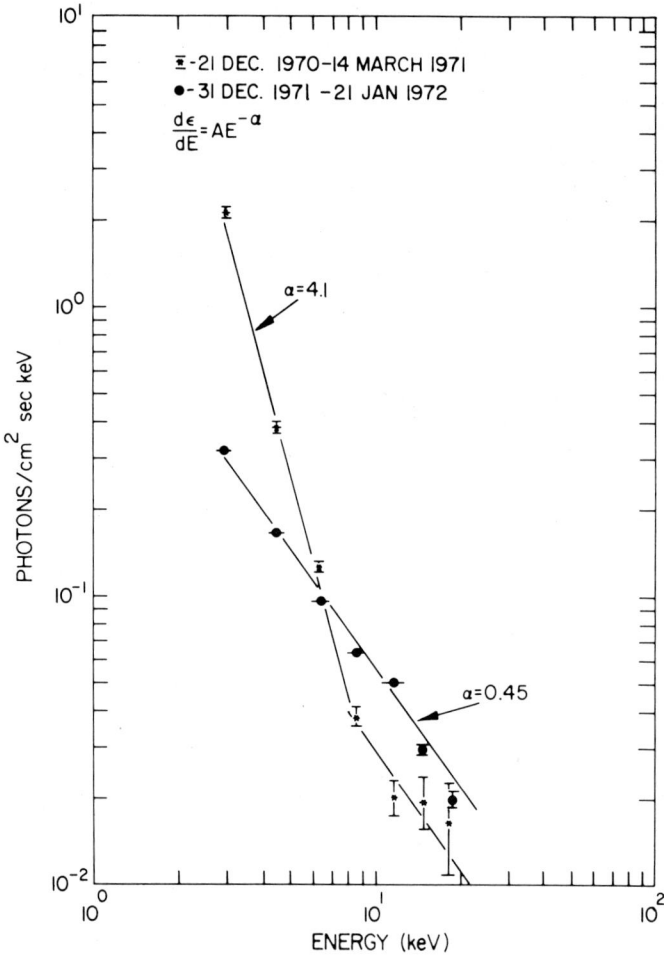

Fig. 16. Cyg. X-1 spectra.

dening and the equivalent width and velocity of the interstellar K line to determine values of 20 $M_\odot$ for the primary, 13 $M_\odot$ for the secondary, 2.2 kpc for the distance, and an inclination angle of 26°. Cherepashchuk (1972) used the absorption line velocities, the HE II emission line velocities with some allowance that the emission region may not belong to the X-ray star, but may lie between the two stars, and the photoelectric observations showing 0.07 mag. changes due to a tidally distorted system. Taking into account limb darkening and the gravity darkening, and assuming that the primary fills its Roche lobe, they determined a primary mass between 10.7

and 22 $M_\odot$, a secondary mass between 7.8 and 17 $M_\odot$, and a distance as large as 5 kpc. They appear to have neglected any interstellar absorption effects and have, therefore, overestimated the distance. Mauder (1973) used the absorption line velocities, the possible distances allowed by the observed reddening (and absence of a bright infrared source which could be produced by an absorbing circumstellar shell), the absence of any substantial reflection effects as demonstrated by the photoelectric observations, the X-ray to visible light energy ratio, and the photoelectric observations. Assuming that the star cannot be any larger than its Roche lobe, he determines a self-consistent set of parameters that gives a distance of 2 kpc, a primary mass of 25 $M_\odot$, and a secondary mass between 6.0 and 7.3 $M_\odot$. Most importantly, Margon et al. (1973) have recently determined the extinction of 50 stars in the field immediately surrounding Cyg X-1 and find $Av/d = (1.3 \pm 0.2)$ kpc^{-1}. Since $Av = 3.3$ for HDE 226868, this leads to a distance estimate for Cyg X-1 of 2.5 ± 0.4 kpc, in agreement with the spectroscopic modulus for a B09 star. I also understand that Kraft will report on a refined determination of distance which agrees with the above result.

In addition, the recent study of Sanduleak 160 reported by Liller and by Hiltner et al. (1973), seems to me to lend much strength to these arguments. Sanduleak No. 160 (13.2 mag. star) had been suggested as the optical counterpart of SMC X-1, the occulting binary X-ray source in the Small Magellanic Clouds. No doubt exists about the identification of this source since the period of the X-ray occultations and of the visible light star coincided. The companion star is reported as B01 according to Webster et al. (1972). Liller (1972) has pointed out that if one derives a value for the absolute visual magnitude of the star following Keenan (1963), one finds -6.2 and -5.6 for the M_v of stars of spectral type B0Ib and B0II. This is in excellent agreement with the value of M_v derived using the distance modulus of SMC of 18.8 mag. and the observed visual magnitude. The absolute value computed in this manner is of 6.0. Thus, for at least this case of SMC X-1, the presence of a strong X-ray source in the system has not altered the spectral appearance of the companion sufficiently to cause substantial errors in determining the mass from spectroscopic data alone. It should be noted that in SMC X-1 the ratio between X-ray and visible light emission is considerably greater than in Cyg X-1. In fact, the heating effects of SK 160 by SMC X-1 are clearly seen in the light curves obtained by Petro et al. (1973) for SK 160 while no such effect has been clearly established for Cyg X-1. Therefore, it appears that the effect of the X-ray flux on the primary is not important for the system containing SMC X-1 and should, therefore, be even less important for the system containing Cyg X-1.

It appears to me that the distance of Cyg X-1 has been clearly established to be greater than 2 kpc, that its luminosity and its mass are those consistent with a B0 supergiant and that, therefore, the value of a mass greater than 6 $M_\odot$ for the X-ray source has also been firmly established.

4.3. Compact Nature of the Source

The rapidity of X-ray intensity variations leaves little doubt that the size of the X-ray emitting region is less than 10^9 cm. This can be used as an argument to imply that the entire star on which the X-ray emission is taking place is compact. This argument depends only on the assumption of an accretion model, and the validity of the Eddington limit.

It seems to me that additional evidence comes from the lack of any observable contribution to the visible light emitted by the system from the 6 $M_\odot$ star. For a main sequence star one could predict that approximately 3% of the light from the system should arise from the object due to its intrinsic luminosity. A similar value would be due to reflection of the light from the primary.

No evidence is observed in the light curve for the presence of such a star and Bolton (1973) has placed a limit of 1% of the total luminosity on its continuum contribution. A weaker upper limit on any line emission arising from the secondary is of order of 15%.

In conclusion, there appears to be strong evidence that Cyg X-1 consists of a binary system containing a B0 supergiant of 20 or 30 $M_\odot$, and a compact source of 6 $M_\odot$. This is the strongest evidence to date for the existence of black holes. I would like to mention that perhaps even more convincing evidence for the existence of such systems will come from the discovery and study of other similar objects. Two, in particular, appear for different reasons as candidates for this search. The first is, of course, the SMC X-1 source. In this case, there appears to be no doubt about the identification and distance of the source and consequently about the mass of the primary. Petro et al point out in a recent paper that from an analysis of the optical photometry and X-ray eclipse duration, the mass ratio is found to be 0.28, $i=79°$, and $P=3\overset{d}{.}89206$. On this basis of a mass of 20 $M_\odot$ for SK 160, the mass of SMC X-1 is 5.6 $M_\odot$. Confirmation of this result requires precise radial velocity measurements which unfortunately have not yet been made. The source is too weak to allow determination of short time variability (0.1–10 keV). However, variability in the time scale of minutes to months is observed.

The source already mentioned previously, Cir X-1, is potentially quite interesting from this point of view. It appears to exhibit the same rapid non-periodic pulsations as Cyg X-1 (Figure 17). Its position is quite well known (0.0002 sq deg) (Figure 18). No bright star appears to be present within the error box brighter than 15 mag. We have some indication of high and low states, however, which may have corresponding visible light variations and might aid in the identification. Also, there appears to be some tentative evidence in the X-rays for a binary occulting behavior. The period would be quite long, ~ 12 days. If an optical counterpart could be found on the basis of either of these effects, among the stars in the error box, further analysis of this system might prove quite interesting, in that although the X-ray emission characteristics closely resemble those of Cyg X-1, the companion star appears to be quite dim and therefore of significantly lower mass than HDE 226868. Under these conditions, if the accreting object is a massive compact object, large radial velocities should be observed for the companion. Most of the mass of the binary system in this case would reside in the compact object.

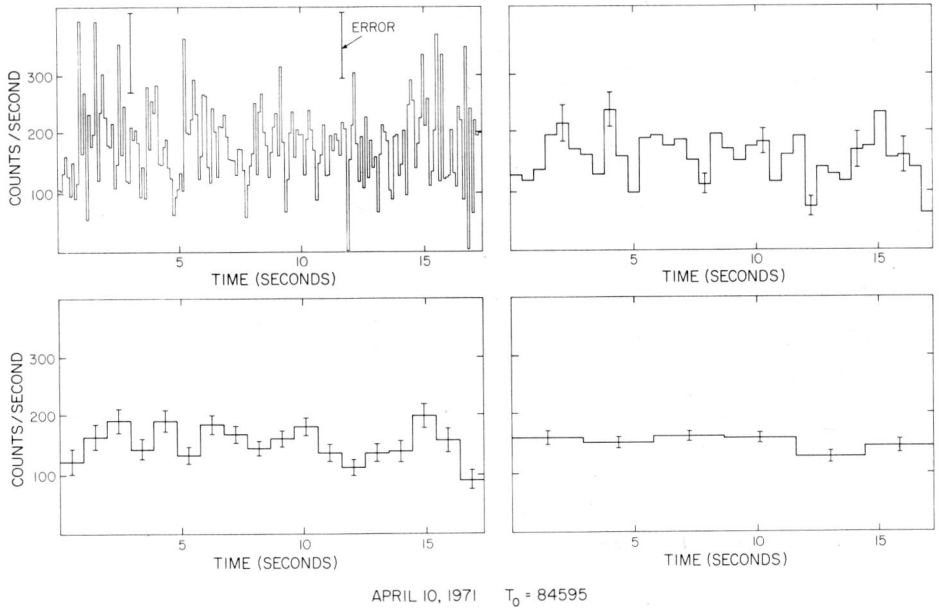

Fig. 17. 3U 1516−56.

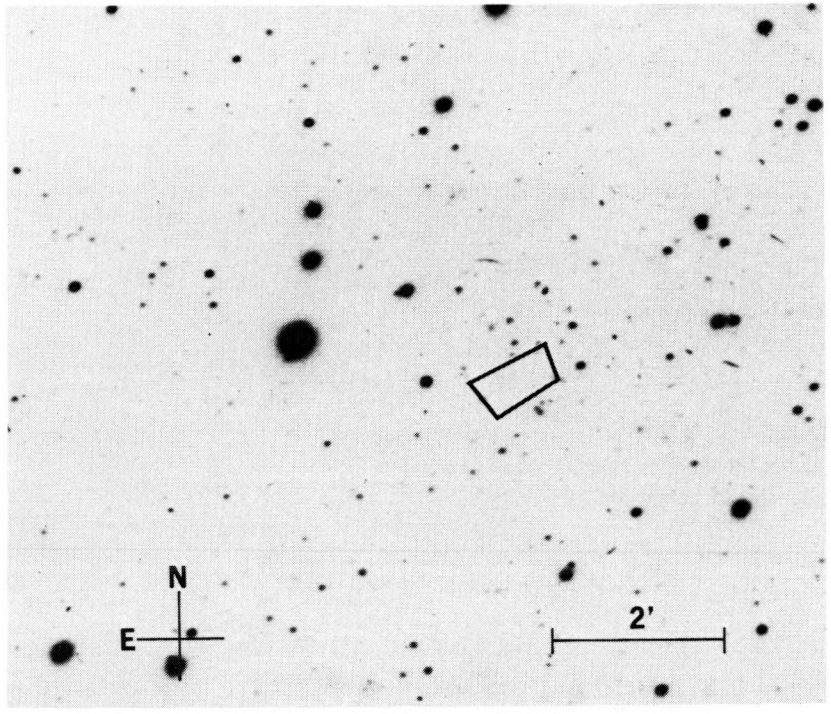

Fig. 18. Location of 3U 1516−56.

5. Other Binary Sources

The discovery of Her X-1, Cyg X-1 and Cen X-3 raises naturally the question of whether most other galactic sources also can be associated with binaries.

At present we have some 6 X-ray sources we believe are associated with binaries on firm grounds. They are: Her X-1, Cen X-3, Cyg X-1, 2U 1700−37, 2U 0900−40, SMX-1. With the exception of Cyg X-1, all other sources exhibit an eclipsing behavior. Cyg X-3 exhibits a periodic 4.8 h intensity variation in X-rays, with a factor of 2 maximum to minimum intensity. A strong variable radio source at the Cyg X-3 location was observed to exhibit enormous flares in September 1972. At the location of the radio source a periodically varying IR source was recently detected by Neugebauer *et al.* exhibiting sharp occultations.

The next several figures show some of the data on known eclipsing X-ray binaries. In particular, Figure 19 shows the data used to demonstrate that 2U 1700−37 eclipses with a 3.4 day period. Data in the upper portion show seven days of data in May 1972 from which the approximate period was obtained. Data in the bottom half of the figure are all of the UHURU data from December 1970 to May 1972

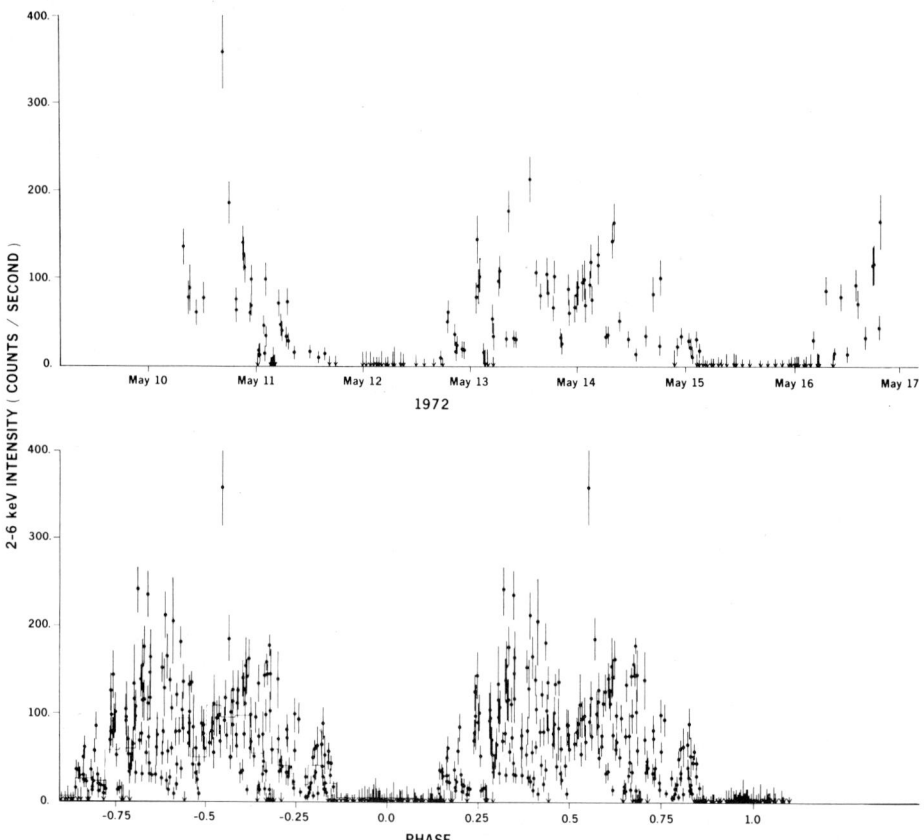

Fig. 19.

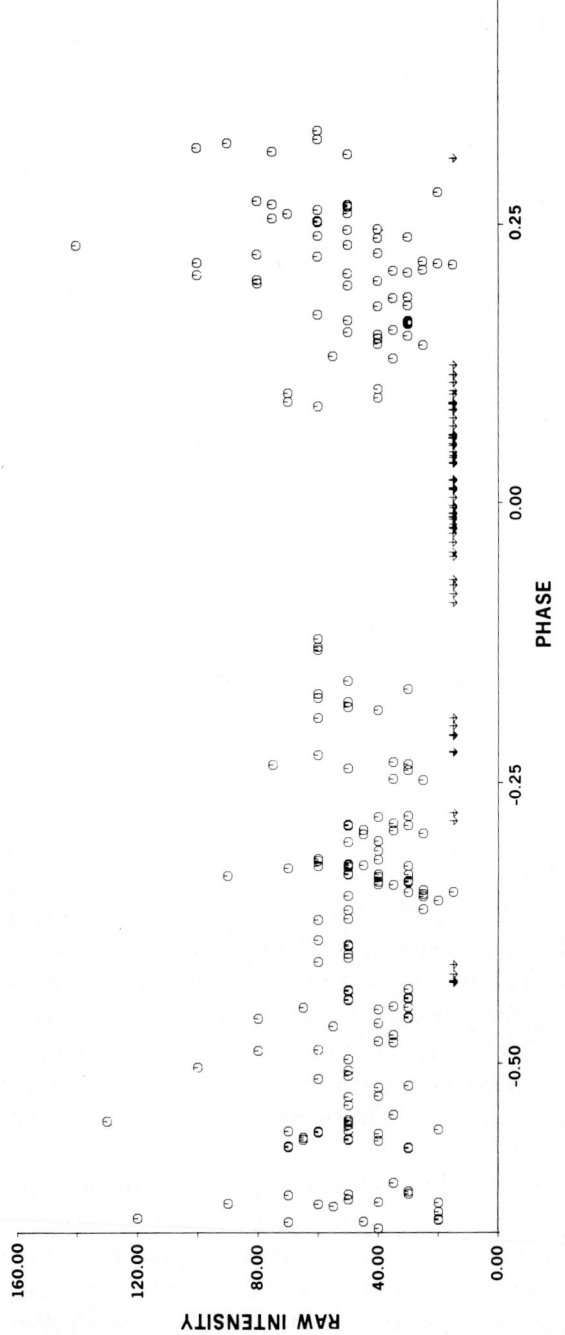

Fig. 20. 172 U 0900−40. Data folded with 8.96 day period (May–June 1972).

folded with a 3.412 day period. Figure 20 shows the UHURU observations of 2U 0900−40 folded with an 8.95 day period. The determination of this period was considerably complicated by the complex, erratic variability that the source undergoes. In particular, times of low intensity at phases other than eclipse are observed, although with no regular pattern. This type of behavior obviously greatly complicates the search for new eclipsing X-ray binaries. Figure 21 shows the UHURU data on the X-ray source in the Small Magellanic Cloud obtained in January and June 1971, showing the source eclipses with a 3.9 day period.

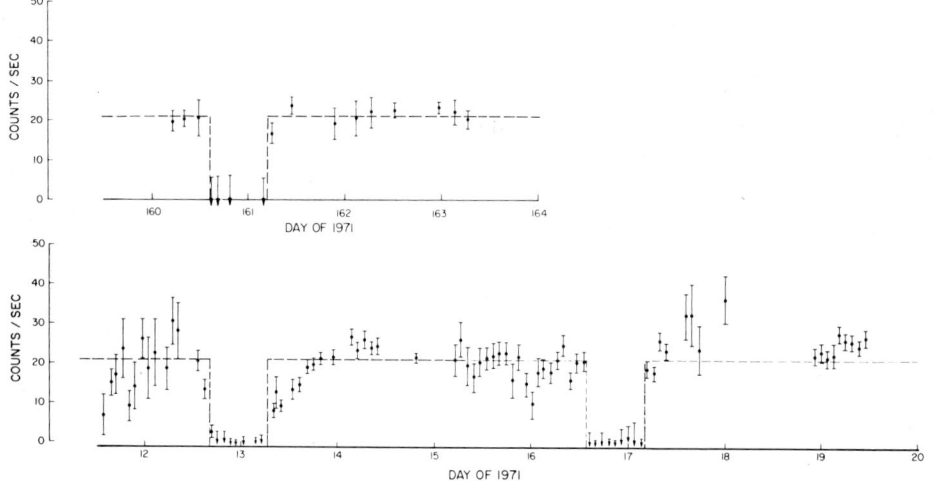

Fig. 21. SMC X-1 (2U 0115−73).

Table III summarizes much of the data on established X-ray binaries. The first column names the 6 sources; the second column gives the binary period (the period for Cyg X-1 is deduced from the behavior of its optical counterpart), the third column describes the short term variability observed in all 6 sources − (SMC X-1 is too weak to detect 0.1 to 1 s pulsations with UHURU), the fourth column names the optical candidate to which the candidate for Cen X-3 should now perhaps be added; the fifth column estimates the distance from the optical data; and the last column gives the peak observed X-ray luminosity from 2–10 keV (the luminosities range from 10^{36} to 10^{38} erg s^{-1}), close to the Eddington limit at which radiation pressure should limit mass flow and thereby luminosity.

Typical counting spectra for these 6 binary sources are shown in Figure 22. Notice the deficit in counts for 5 of the 6 sources in the lowest energy channels. The only source without a substantial low energy cutoff is the only one which does not eclipse − Cyg X-1. We interpret this as being caused by the presence of much material in the orbital plane of the mass transfer binary systems. Systems which eclipse are observed at inclinations near the orbital plane, where much of the material of the system presumably lies, thus giving rise to the observed absorption. Cyg X-1 is presumed to be observed from above the orbital plane and hence little obscuring material is

TABLE III

CHARACTERISTICS OF X-RAY BINARIES

Source	Binary Period (Days)	Short Term Variability	Optical Candidate	Distance (kpc)	Peak Luminosity (2-10 keV ergs/sec)
Cyg X-1 (2U1956+35)	5.600 ± 0.003 from optical. Not observed in X-ray.	Quasi-periodic pulsations as short as 50 millisec.	HDE226868	2	1×10^{37} before transition 3×10^{36} after transition.
Cen X-3 (2U1119-60)	2.08712 ± 0.00004	4.842 sec pulsations	None	?	?
Her X-1 (2U1702+35)	1.700167 ± 0.000006	1.23782 sec pulsations	HZ Her	5.8	1×10^{37}
2U1700-37	3.412 ± 0.002	Non-periodic pulsations as short as 0.1 sec	HD153919	1.7	3×10^{36}
2U0900-40 (GX263+3)	8.96 ± 0.05	Non-periodic pulsations on times of secs	HD77581	1.3	4×10^{36}
SMC X-1 (2U115-73)	3.8927 ± 0.0010	Non-periodic pulsations on times of mins	Sanduleak 160	61	3×10^{38}

along the line of sight. The fact that intensity variation appear to be a prevalent characteristic of the X-ray sources in binary systems makes it very difficult to establish the existence of eclipsing behavior. It is also clear that only a few of the sources even if they are in binary systems will have appropriate orbital inclindations to allow us to observe eclipses.

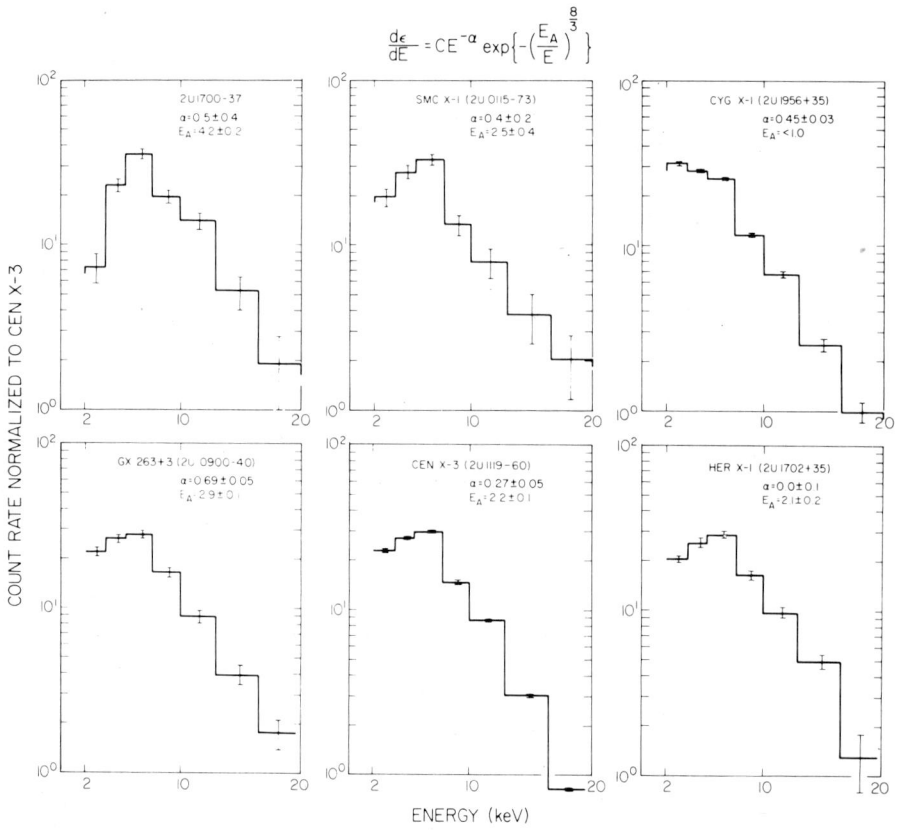

Fig. 22. Binary source spectra.

We have, therefore attempted to establish other criteria to aid us in determining how many of the strong galactic sources could be binaries.

The criteria of large intensity variations and substantial low energy cutoffs were used in selecting the 9 candidates for first study as possible new eclipsing binaries. While no new binaries have been observed definitely, the data in Figure 23 indicate the situation with Cir X-1 (3U 1516−56). The data on this source were examined for possible periodicities less than 15 days and a period of 12.29 days was obtained. However, due to the erratic behavior of the source and the fact that it has not been possible to observe the source continuously for several periods, we regard this result as tentative. Figure 23 shows the data folded with a 12.3 day period and we see there is about $\frac{3}{4}$ of a day with no high intensity sightings. As we earlier mentioned with

2U 0900−40 there are many additional low points, as well as a large scatter, showing the extreme variability of this source. All of this contributes to the difficulty in obtaining a period and thereby confirmation as a binary. Given the very small location error box for this source, and its intrinsic interest, we hope that optical astronomers may succeed in finding an optical counterpart.

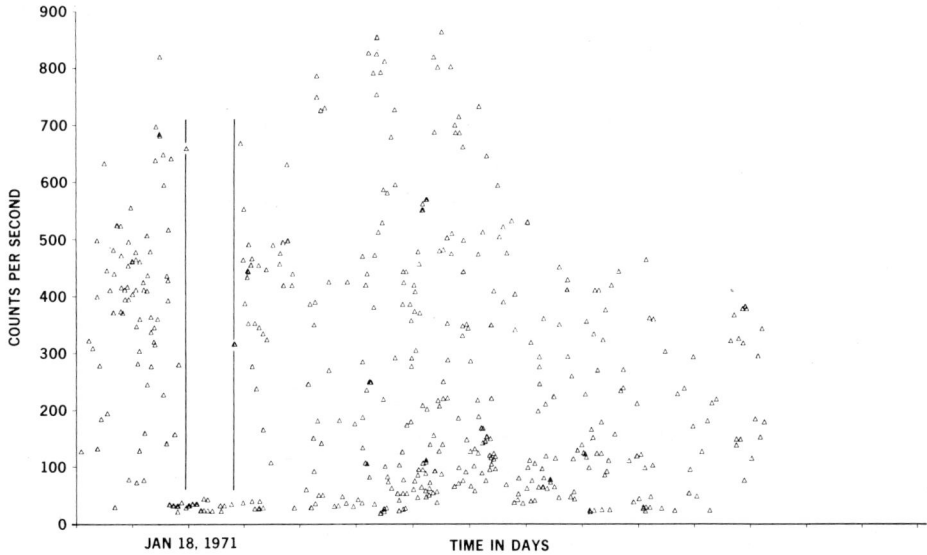

Fig. 23. Cir X-1, data folded with 12.29 days.

In view of the limited success of this approach, we have begun a systematic search for short and long term variability, adopting the point of view that compactness is a necessary, if not sufficient, criterion for establishing the binary nature of the sources. Although this work is just beginning, I would like to report some preliminary results that have been obtained by our group and primarily by Christine Forman Jones.

5.1. Long term variability

In the first table (Table IV) a compilation of the day-to-day variability of the strongest among the known 161 X-ray sources that comprise the 3U Catalog is given (Giacconi et al., 1973). We have divided the sources into two intensity classes, the 44 sources greater than 35 counts s^{-1} and the 10 sources between 20 and 35 counts s^{-1} (sources fainter than 20 counts s^{-1} have insufficient counting statistics to be considered for this purpose). Before proceeding to examine this sample for variability, we eliminate known extragalactic sources (NGC 1275 and Virgo Cluster), known supernova remnants (the Crab and Cas A), and the known extended source at our Galactic Center. This leaves 40 and 9 sources in the two categories, respectively. Of the 40 sources brighter than 35 counts s^{-1}, 36 are found to vary in average intensity on different days (1 day to several months apart) in the 3U Catalog. Additional study has shown that 3U 1636−53 and 3U 1728−24 also vary in intensity, leaving only 3U 1735−44 (for which we have parts of 4 days of data) and 3U 1822−00 at 37 counts s^{-1} as not variable. Based on past experience, 3U 1735−44 might be found to vary as more data become available, although another possibility is that this source could correspond to a supernova remnant with weak, extended, non-thermal

TABLE IV

DAY-TO-DAY VARIABILITY OF 3U SOURCES

161 SOURCES IN 3U CATALOG

	I > 35 COUNTS/SEC	35 ≥ I ≥ 20
TOTAL SOURCES	44	10
KNOWN EXTRAGALACTIC	1 (NGC 1275)	1 (VIRGO)
KNOWN SUPERNOVA	2 (CRAB, CAS A)	
KNOWN EXTENDED	1 (GAL. CENTER)	
TOTAL REMAINING	40	9
VARY IN 3U CATALOG	36	3
VARY FROM FURTHER STUDY	2	
NOT PRESENTLY KNOWN TO VARY	2 (3U 1735-44 <I>=210 3U 1822-00 <I>=27)	6
KNOWN BINARIES	6*	1
CANDIDATES UNDER STUDY AS POSSIBLE ECLIPSING BINARIES	9	

*INCLUDES CYG X-1 WHICH DOES NOT ECLIPSE IN X-RAYS AND CYG X-3

radio emission yet undetected. As for 3U 1822−00 a factor of two variability could easily have gone undetected given the source intensity of 37 counts s^{-1}, as is also suggested by the absence of detected variability in 6 of the 9 sources with intensity between 20 and 35 counts s^{-1}. This is apparently the boundary at which the UHURU instrument becomes limited in sensitivity. Thus, our data are consistent with the view that all sources exhibit large variations of intensity on the scale of days. These results have been obtained by comparing the average intensity of the sources from one sighting to the next.

5.2. SHORT TERM VARIABILITY

We can now examine each individual sighting in order to study the short time scale variability (0.1 to 1 s) of the more intense galactic sources. Table V is a table summarizing the status of this survey. The table starts with the 2 periodically pulsing sources – Cen X-3 and Her X-1, and 4 sources already reported by us in the literature as pulsating but with no evidence for regular, persistent periodicity. Then, we list 8 sources newly found to pulsate and 4 more which probably pulsate. The results

TABLE V
SHORT* TIME SCALE VARIABILITY OF 3U SOURCES

PERIODIC VARIATIONS	PROBABLY PULSATE
1. CEN X-3 (3U1118-60)	
	15. CYGNUS X-3 (3U2030+40)
2. HER X-1 (3U1653+35)	
	16. GX13+1 (3U1811-17)
PREVIOUSLY REPORTED AS PULSATING	17. CYGNUS X-2 (3U2142+38)
3. CYG X-1 (3U1956+35)	18. 3U1658-48
4. 3U1700-37	
	NO EVIDENCE FOR LARGE PULSATIONS
5. CIRC X-1 (3U1516-56)	
6. GX263+3 (3U0900-40)	
	19. GX17+2 (3U1813-14)
NEWLY FOUND TO PULSATE	20. GX9+9 (3U1728-16)
7. 3U1636-53	21. 3U1630-47
8. GX9+1 (3U1758-20)	22. 3U1705-44
9. GX349+2 (3U1702-36)	23. SCO X-1 (3U1617-15) - shows ~5% variations in 1 SEC
10. SERPENS X-1 (3U1837+04)	
11. 3U1820-30	STUDIED BUT TOO WEAK TO DRAW CONCLUSIONS
12. GX3+1 (3U1744-26)	
13. GX5-1 (3U1758-25)	24. 3U1702-42
14. GX340+0 (3U1652-45)	25. 3U1822-00
	26. 3U1727-33
	27. 3U0115+63
	28. SMC X-1 (3U0115-73)

*SHORT – → TYPICALLY LESS THAN 1 SECOND AND OFTEN ON TIMES OF 0.1 SECOND.

are obtained by means of a X^2 analysis of the intensity fit to individual passes of durations from 2 to 100 s and refer to non-periodic variations. Not all passes exhibit a statistically significant X^2, but for the sources numbered 7 to 14 there are sufficient passes with large enough X^2 to conclude the sources do pulsate (at least some of the time). For sources 15–18, not enough data have yet been analyzed, but preliminary indications strongly suggest the existence of pulsations. In the case of Cyg X-2, variations of at least 25% in less than 1 s are observed on one occasion.

For the other sources, 7 to 18 in the table, variability is often observed on times of 0.1 s – the best time resolution available with UHURU and with amplitudes that are consistent with 50% changes in intensity. What we have not yet done is determine quantitatively the characteristic time and amplitude of the variability – if such categorization should prove significant – or the fraction of the time that the source is active. Sources 19–23 in the table were studied in a number of passes and have not yet demonstrated large amplitude, short time scale variations, although all have been observed to vary on times of days or less.

The last 5 objects in the table are sources whose intensity is too low to allow us to draw any conclusions from our analysis. Although this survey has not yet been completed and we cannot yet give a quantitative description of the characteristic time scales and amplitude of the variation, yet the above results allow us to conclude that essentially all of the strong galactic X-ray sources with the exception of supernova remnants are variable and many of them appear to vary on times less than 1 s, indicating source regions of 10^{10} cm or less.

It may well be that the variety of X-ray behavior we observe is caused by combinations of the various parameters which a binary system with a compact secondary may exhibit, starting with the nature of the secondary, its magnetic field, the mass transfer rate, and the inclination angle to the observer, to name a few. We hope that the survey for eclipsing X-ray sources, the study of the X-ray light curves of the 40 or so brightest sources, the search for short time scale variability, and the searches for optical counterparts presently underway, will contribute to our understanding of the physical processes taking place at the source and perhaps allow us to answer, in at least a statistical sense, the question of the binary nature of all of the galactic X-ray sources and to begin to understand the complicated variety of behavior exhibited. In the near future with the advent of orbiting X-ray telescopes, such as the one planned for the HEAO-B mission of NASA, it will become possible to extend the study of sources of 10^{36}–10^{37} erg s^{-1} intrinsic luminosity to 30 Mpc. It is our hope that this will greatly expand the number of sources we will be able to investigate in detail, and make it possible to give a substantial contribution to our understanding of the properties of objects near the end point of stellar evolution.

Note added in proof. It has recently been reported by a group at Goddard Space Flight Center (Rothschild *et al.*, 1973) that during a rocket experiment in which Cyg X-1 was observed with high time resolution (~ 300 μs) for a period of 100 s, large amplitude variations occurring in times of the order of 1 ms were observed. This report confirms the analysis performed by Oda *et al.* (1973) on the rocket experiments by the MIT group (Rappaport *et al.*, 1971b). This finding further strengthened the conclusion about the compact nature of the Cyg X-1 source.

Acknowledgements

Many of the X-ray results that I have discussed in this paper were obtained by the UHURU group at American Science and Engineering, now at the Center for Astrophysics (SAO/HCO) at Harvard. Among this group are Harvey Tananbaum, Ethan Schreier, William Forman, and Christine Jones Forman, to whom I would like to express my gratutide for their assistance in the preparation of this material. I would also like to thank Wallace Tucker and George Field for their helpful discussions.

References

Baker, K. and Ruffini, R.: 1973, 'Theoretical Implications of the First Derivative of the Periods in Binary X-rays Sources', preprint.
Bolton, C. T.: 1972, *Nature* **235**, 271.
Bolton, C. T.: 1973a, 'Dimensions of the Binary System HDE 226868 = Cygnus X-1', preprint.
Bolton, C. T.: 1973b, private communications.
Braes, L. and Miley, G. K.: 1971, *Nature* **232**, 246.
Burbidge, E., Lynds, C., and Stockton, A.: 1967, *Astrophys. J.* **150**, L. 95.
Cherepashchuk, A. M., Lyutiy, V. M., and Sunyaev, R. A.: 1973, *Soviet Astron. AJ* **50**, in press.
Colgate, S. and White, R. H.: 1966, *Astrophys. J.* **142**, 626.
Forman, W., Jones, C. A., and Liller, W.: 1972, *Astrophys. J. Letters* **178**, L103.
Giacconi, R.: 1967, in L. Perek (ed.), *Highlights of Astronomy*, D. Reidel Publ. Co., pp. 192–201.
Giacconi, R., Gursky, H., Kellogg, E., Levinson, R., Schreier, E., and Tananbaum, H.: 1973a, *Astrophys. J.* **184**, 1.
Giacconi, R., Murray, S., Gursky, H., Kellogg, E., Schreier, E., Matilsky, T., Koch, D., and Tananbaum, H.: 1973b, 'The UHURU Catalog of X-ray Sources', in press.
Hjellming, R. M. and Wade, C. M.: 1971, *Astrophys. J. Letters* **168**, L21.
Holt, S., Boldt, E., Schwartz, D., Serlemitsos, P., and Bleach, R.: 1971, *Astrophys. J. Letters* **166**, L65.
Keenan, P. C.: 1963, *Stars and Stellar Systems* **3**, 92.
Kraft, R.: 1964, *Astrophys. J.* **139**, 457.
Lamb, F. K., Pethick, C. J., and Pines, D.: 1973 'A Model for Compact X-Ray Sources: Accretion by Rotating Magnetic Stars', preprint.
Liller, W.: 1973, *Astrophys. J. Letters* **184**, L37.
Margon, B., Bowyer, S., and Stone, R.: 1973, 'On the Distance to Cygnus X-1', preprint.
Mauder, H.: 1973, 'On the Mass Limit of the X-Ray Source Cyg X-1', preprint.
Oda, M., Gorenstein, P., Gursky, H., Kellogg, E., Schreier, E., Tananbaum, H., and Giacconi, R.: 1971, *Astrophys. J. Letters* **166**, L1.
Oda, M., Matsuoka, M., Miyamoto, S., Ogawara, Y., and Takagishi, K.: 1973, 'X-Ray Pulsation of Cyg X-1', paper presented at the *Fall Meeting of the Astronomical Society of Japan*.
Ostriker, J. P. and Davidson, K.: 1973, in H. Bradt and R. Giacconi (eds.), 'X- and Gamma-Ray Astronomy', *IAU Symp.* **55**, 143.
Paczyński, B.: 1971, *Ann. Rev. Astron. Astrophys.* **9**, 183.
Paczyński, B.: 1972, 'Comments at the Texas Symposium for Relativistic Astrophysics'.
Petro, L., Feldman, F., and Hiltner, W. A.: 1973, 'Optical Observations of Sanduleak 160', preprint.
Prendergast, K. and Burbidge, B.: 1968, *Astrophys. J.* **151**, L83.
Pringle, J. E. and Rees, M. J.: 1972, *Astron. Astrophys*, **21**, 1.
Rappaport, S., Zaumen, W., and Doxsey, R.: 1971a, *Astrophys. J. Letters* **168**, L17.
Rappaport, S., Doxsey, R., and Zaumen, W.: 1971b, *Astrophys. J. Letters* **168**, L43.
Rotschild, R., Boldt, E., Holt, S., and Serlemitsos, P.: 1973, 'Temporal and Spectral Studies of Her X-1, Cyg X-1 and Cyg X-3', paper presented at the *Joint Meeting between the Div. of Cosmic Physics (APS) and the High Energy Astrophysics Div. (AAS)*, Tucson, Arizona.
Sandage, A., Osmer, P., Giacconi, R., Gorenstein, P., Gursky, H., Waters, J., Bradt, H., Garmire, G., Sreekantan, B., Oda, M., Osawa, K., and Jugaku, J.: 1966, *Astrophys. J.* **146**, 316.
Schreier, E., Gursky, H., Kellogg, E., Tananbaum, H., and Giacconi, R.: 1971, *Astrophys. J. Letters* **170**, L21.
Shakura, N. I. and Sunyaev, R. A.: 1973, *Astron. Astrophys.* **24**, 337.
Shulman, S., Fritz, G., Meekins, J., and Friedman, H.: 1971, *Astrophys. J. Letters* **168**, L49.

Shvartsman, V. F.: 1971, *Astron. Zh.* **48**, 438; *Soviet Astron*, **15**, 343.
Trimble, V., Rose, W. K., and Weber, J., 1973, *Monthly Notices Roy. Astron. Soc.* **162**,
Van den Heuvel, E. P. J. and Heise, J.: 1972, 'Centaurus X-3 Possible Reactivation of an Old Neutron Star by Mass Exchange in a Close Binary', preprint.
Walborn, N. R.: 1973, *Astrophys. J. Letters* **179**, L123.
Webster, B., Martin, W., Feast, M., and Andrews, P.: 1973, *Nature Phys. Sci.* **240**, 183.
Webster, L. and Murdin, P.: 1972, *Nature* **235**, 37.
Wheeler, J. A. and Ruffini, R.: 1971 (January), *Physics Today*, p. 30.
Zel'dovich, Ya. B. and Novikov, I. D.: 1964, *Soviet Phys. Doklady* **9**, 246.
Zel'dovich, Ya. B. and Novikov, I. D.: 1971, *Relativistic Astrophysics*, Vol. 1, Univ. of Chicago Press, Chapter 13.

ON THE DISTANCE TO CYGNUS X-1 (HDE 226868)*

JESSE BREGMAN, DENNIS BUTLER, EDWARD KEMPER, ALAN KOSKI
R. P. KRAFT, and R. P. S. STONE**

*Lick Observatory, Board of Studies in Astronomy and Astrophysics,
University of California, Santa Cruz, Calif., U.S.A.*

Abstract. From U, B, V photometry of 104 stars in a field of radius $\sim 30'$ centered on the X-ray binary star Cyg X-1 (HDE 226868), we have studied the color excess $E(B-V)$ as a function of distance. Spectral types were observed *de novo* for 42 of these stars. We conclude that HDE 226868 cannot be nearer than 1 kpc, and is probably at a distance of 2.5 kpc or more. The primary component is therefore a luminous OB star of mass $\sim 30\,\mathfrak{M}_\odot$, and the X-ray component has a minimum mass $\sim 6\,\mathfrak{M}_\odot$.

---

* Contributions from the Lick Observatory No. 385, *Astrophys. J. Letters* **185**, L117, November 15, 1973.
** Supported in part by NSF Science Departmental Improvement Grant GU-3162.

ORIGIN OF COSMIC RAYS, ATOMIC NUCLEI AND PULSARS IN EXPLOSIONS OF MASSIVE STARS

W. DAVID ARNETT

Dept. of Astronomy and Physics, University of Texas at Austin, Austin, Tex. 78712, U.S.A.

Abstract. Explosions of massive stars $(8 \lesssim M/M_\odot \lesssim 70)$ are examined as the source of galactic cosmic rays. Detailed nucleosynthetic and evolutionary calculations suggest that these massive stars produce the heavy elements (carbon and above) in their proper relative abundances. This is particularly significant because lower mass stars (in particular the 4–8 $M_\odot$ range) are not able to produce the observed abundances of C and O relative to the iron peak. A small ($\sim 1.4\ M_\odot$) dense remnant star (a neutron star) left after the explosion may provide a location for an electromagnetic acceleration mechanism. Those abundance ratios which can now be predicted (He, C, O, Ne, Mg) for the material to be accelerated by the pulsar give a reasonable match to the observed cosmic ray data. The conditions at the outer edge of the remnant and the inner edge of the ejected material may be appropriate for an **r**-process to occur; the high Z cosmic rays seem to show an enrichment of **r**-process material. It appears that these stars may be the astrophysical source for the galactic cosmic rays. The questions of rotation and black hole formation were discussed. It appears that the most straight-forward result of evolution of a close massive binary is a massive star and a neutron star in a low eccentricity orbit, in agreement with observation.

ON THE PROBLEM OF DETECTION OF ISOLATED BLACK HOLES

V. F. SHVARTSMAN

Special Astrophysical Observatory, U.S.S.R. Academy of Sciences, St. Zelenchukskaya, Stravropolsky Kraj, 357140 U.S.S.R.

Abstract. The interstellar plasma accreted by isolated black holes should be heated during the infall process and radiate in the magnetic fields connected with the infalling material. A portion of the radiation in the optical range is likely to have a luminosity of order 10^{27}–10^{30} erg s^{-1}. In *SAO U.S.S.R. AS* an experiment is being carried out to observe isolated black holes as faint optical stars without spectral lines. So far the results of the experiment are negative.

References

Shvartsman, V. F.: 1970, preprint, IAM 42, *Soviet Astron. AJ* **15**, 377, 1971.
Shvartsman, V. F.: 1973, *Commun. SAO U.S.S.R. AS*, in press.

BLACK HOLES IN THE EARLY UNIVERSE*

BERNARD J. CARR and STEPHEN W. HAWKING

Institute of Astronomy, Cambridge, England

Abstract. The existence of galaxies indicates that the early universe must have been inhomogeneous and might have been highly chaotic. This could have lead to regions of the size of the particle horizon undergoing gravitational collapse to produce black holes with initial masses from 10^{-5} g upwards. Radiation pressure in the early Universe would cause these black holes to grow by accretion. However, despite previous expectations, this accretion would not be very much unless the initial conditions of the Universe were arranged in a special and a causal manner. Observations indicate that, at the most, only a small fraction of the matter in the early Universe can have undergone gravitational collapse.

* To be published.

QUANTUM ASPECTS OF ACCRETION ONTO BLACK HOLES IN THE EARLY UNIVERSE*

S. W. HAWKING and G. W. GIBBONS

Institute of Astronomy, Cambridge, England

Abstract. It is argued that for black holes of mass less than roughly 10^{17} g formed in the early Universe, the accretion process must be calculated using quantum mechanics. It is also argued that primordial black holes must be more massive than 10^{17} g if they are to possess charge.

* To be published.

SURFACE COMPOSITION OF NEUTRON STARS THAT ARE ACCRETING MATTER*

S. TSURUTA, R. RAMATY, and G. BÖRNER

NASA, Goddard Space Flight Center, U.S.A.

Abstract. Surface composition of a neutron star in the absence of accretion is pure iron (Tsuruta, 1964). We considered two cases of accretion. (a) The accretion of inter-stellar matter: Here we expect the presence of a significant amount of various elements between proton and iron, though iron is still the most dominant. (b) The accretion as the energy source of an X-ray source in a binary system. Here it is most likely that the incoming flux is sufficiently high (non-steady state) and protons accumulate on the accreting region of the surface.

Reference

Tsuruta, S.: 1964, Ph.D. Thesis, Columbia University.

* To be published.

THE MAGNETIC FIELDS OF PULSARS

D. M. SEDRAKIAN

Dept. of Physics, State University of Erevan, U.S.S.R.

Abstract. Two generation mechanisms of magnetic fields in pulsars are considered.

If the temperature of a star is more than 10^8 K, the star consists of a normal fluid of neutrons, protons and electrons. Because the angular velocity of pulsars is not constant $d\Omega/dt \neq 0$, inertia effects can occur, and generate magnetic fields through the relative motion of charged particles with different masses. The kinematic viscosity of electrons is 30 times larger than that of protons; hence electrons move with the crust, but the proton-neutron fluid will move relative to the electrons. The magnetic momentum can be calculated by the following formula

$$\mathbf{M} = \frac{4\pi M_{\text{eff}} \cdot R^5 \sigma}{15ce} \frac{d\Omega}{dt}, \tag{1}$$

where $M_{\text{eff}} = M_p + M_n(N_n/N_p)$, R = radius of the star, σ = conductivity. For typical neutron stars we have $d\Omega/dt \sim 10^{-8}$ s^{-2}, $R \sim 10^6$ cm, $\sigma \sim 10^{29}$ s^{-1} and we get a magnetic field of the order of 10^{10} G.

The second mechanism appears when the neutrons and protons are a rotating superfluid, while the electrons are still normal. This produces an additional motion of the protons relative to the electrons. The magnet moment can be calculated by the formula

$$M = \frac{3\mu_B}{4} N_p \ln b/a, \tag{2}$$

where μ_B is Bohr's magneton for protons, N_p is the total number of protons in a neutron star and $\ln b/a \sim 10$, where a is the radius of the core of the vortex, b is a radius of vortex. For neutron Stars $N_p \simeq 10^{54}$ and we get the magnetic field of the order of 10^{14} G.

Reference

Landau, L. D. and Lifshitz, E. M.: 1957, *Electrodynamics of Continuous Media* Addison-Wesley, Reading, Mass., U.S.A.

A CLASS OF SOLUTIONS OF EINSTEIN-MAXWELL EQUATIONS WITH THE COSMOLOGICAL CONSTANT

J. F. PLEBAŃSKI

The Institute of Theoretical Physics of the University of Warsaw, Hoża 69, Warsaw, Poland

Abstract. Working in the signature $(+++-)$ and units such that $G=1=c$, it was found a solution of Einstein-Maxwell equations with λ (without current and pseudo-current). In real coordinates $x^\mu = (p, \sigma, q, \tau)$ the solutions is:

$$\omega =: \tfrac{1}{2}(f_{\mu\nu} + \check{f}_{\mu\nu})\, dx^\mu \wedge dx^\nu = -d\left\{\frac{e_0 + ig_0}{q+ip}(d\tau - ipq\, d\sigma)\right\}, \tag{1}$$

$$ds^2 =: \frac{p^2+q^2}{P}\, dp^2 + \frac{P}{p^2+q^2}(d\tau + q^2\, d\sigma)^2 +$$

$$+ \frac{p^2+q^2}{Q}\, dq^2 - \frac{Q}{p^2+q^2}(d\tau - p^2\, d\sigma)^2, \tag{2}$$

where

$$P =: b - g_0^2 + 2n_0 p - \varepsilon p^2 - \tfrac{2}{3}p^4,$$
$$Q =: b + e_0^2 - 2m_0 q + \varepsilon q^2 - \tfrac{2}{3}q^4, \tag{3}$$

$[\check{f}^{\mu\nu} =: (i/2\sqrt{-g})\, \varepsilon^{\mu\nu\varrho\sigma} f_{\varrho\sigma}$ is pure imaginary; in (1) 'd' denotes the external differential]. Not all constants m_0, n_0, e_0, g_0, b, ε, λ are physically significant: by re-scaling coordinates ε can be made equal to $+1, 0,$ or -1. The solution is of the type D: the double Debever-Penrose vectors

$$\text{ff} \qquad \pm k^{(\pm)}_\mu dx^\mu =: d\left(\tau \pm \int \frac{q^2\, dq}{Q}\right) - p^2\, d\left(\sigma \mp \int \frac{dq}{Q}\right) \tag{4}$$

have the common complex expansion $Z = (q+ip)^{-1}$. Among $C^{(a)}$'s only $C^{(3)}$ given by:

$$C^{(3)} = \frac{-2}{(q+ip)^2}\left\{\frac{m_0 + in_0}{q+ip} - \frac{e_0^2 + g_0^2}{q^2+p^2}\right\} \tag{5}$$

is in general $\neq 0$. The invariants of the electromagnetic field are:

$$F =: \tfrac{1}{4} f_{\mu\nu} f^{\mu\nu} + \tfrac{1}{4} f_{\mu\nu} \check{f}^{\mu\nu} = -\frac{1}{2}\frac{(e_0+ig_0)^2}{(q+ip)^4}. \tag{6}$$

The constants contained in (1)–(6) have the interpretation of: (1) e_0 and g_0 are the electric and magnetic monopoles charges respectively, (2) m_0 and n_0 are the mass and NUT parameters (3) b is related to the Kerr constant (4) λ is cosmologic constant (5) the sign ε in the sub-family of solutions which contains Kerr metric is equal to $+1$. [With $\varepsilon=1$, $\lambda=0$ the result described above amounts to the charged Kerr-Newman-NUT metric generalized by the presence of the magnetic monopole; here $b=g_0^2-n_0^2+a_0^2$ where a_0 is the Kerr constant.]

For a test particle of mass Δm which carries electric and magnetic charges Δe, Δg the Hamilton-Jacobi equation is separable: The solution of this equation is:

$$W=C_\tau\cdot\tau+C_\sigma\cdot\sigma+\varepsilon_1\int\frac{dp}{\sqrt{P}}\left[C_0-(\Delta m)^2\,p^2-\frac{1}{P}(p^2C_\tau-p\Delta_g+C_\sigma)\right]^{1/2}$$

$$+\varepsilon_2\int\frac{dq}{\sqrt{Q}}\left[-C_0-(\Delta m)^2\,q^2+\frac{1}{Q}(q^2C_\tau-q\Delta_e-C_\sigma)\right]^{1/2} \qquad (7)$$

where

$$\varepsilon_1^2=1=\varepsilon_2^2, \qquad \Delta_e+i\Delta_g=:(\Delta e-i\Delta g)(e_0+ig_0). \qquad (8)$$

and C_τ, C_σ, C_0 are the separation constants.

Working together with M. Demiański we generalized these results as follows: we have a solution of Maxwell-Einstein equations with λ described by:

$$\omega=d\left\{\frac{e+ig}{1-ipq}(q\,d\tau+ip\,d\sigma)\right\} \qquad (9)$$

$$ds^2=\frac{1}{(p+q)^2}\cdot\left\{\frac{1+(pq)^2}{P}dp^2+\frac{P}{1+(pq)^2}(d\sigma+q^2\,d\tau)^2+\right.$$

$$\left.\neq\frac{1+(pq)^2}{Q}dq^2\equiv\frac{Q}{1\neq j6ql^2}jdJ\equiv 6^2\,d\sigma l^2\right\} \qquad (10)$$

$$P=:\left(\frac{-\lambda}{6}-g^2+\gamma\right)+2np-\varepsilon p^2+2mp^3+\left(\frac{-\lambda}{6}-e^2-\gamma\right)p^4$$

$$Q=:\left(\frac{-\lambda}{6}+g^2-\gamma\right)+2nq+\varepsilon q^2+2mq^3+\left(\frac{-\lambda}{6}+e^2+\gamma\right)q^4 \qquad (11)$$

endowed in continuous constants m, n, e, g, ε, γ, λ. This is also a solution of the type D with twisting double Debever-Penrose directions.

We have here:

$$C^{(3)}=:2(m+in)\left(\frac{p+q}{1-ipq}\right)^3-2(e^2+g^2)\left(\frac{p+q}{1-ipq}\right)^3\frac{p-q}{1+ipq} \qquad (12)$$

$$F=:-\tfrac{1}{2}(e+ig)^2\left(\frac{p+q}{1-ipq}\right)^4. \qquad (13)$$

The transformation $q\to -1/q$, then $(p,q)\to(1/e)(p,q)$, $\tau\to e\tau$, $\sigma\to e^3\sigma$; $P\to e^4 P$, $Q\to e^4 Q$, $e+ig\to e^{-2}(e_0+ig_0)$, $m+in\to e^{-3}(m_0+in_0)$, $\varepsilon\to e^{-2}\varepsilon$, $\gamma\to e^{-4}b+(\lambda/6)$, $\lambda\to\lambda$ yields in the limit $e\to\infty$ the solution previously described by (1)–(6). Another contraction: $(p,q,\sigma,\tau)\to e^{-1}(p,q,\sigma,\tau)$, $n\to ne$, $\varepsilon\to\varepsilon e^2$, $m\to me^3$, $e+ig\to(e_0+ig_0)e^2$, $\gamma\to\gamma+e^4g^2$,

$\lambda \to \lambda$, and then $e \to \infty$ brings the solution to the Kinerseley-Walker family of solutions.

The solution described by (9)–(13) in general is not separable. Constants e, g, m, n are related to electric and magnetic charges, mass and NUT parameters; λ is the cosmological constant; it is conjected that 'kinematical constants' γ and ε are related to uniform acceleration and rotation parameters (γ in contractions corresponds to the Kerr constant).

NEW SOLUTIONS OF EINSTEIN EQUATIONS REPRESENTING SPINNING MASSES

HUMITAKA SATO

Research Institute for Fundamental Physics, Kyoto University, Japan

and

AKIRA TOMIMATSU

Research Institue for Theoretical Physics, Hiroshima University, Takehara, Hiroshima-ken, Japan

Abstract. We found new, stationary axisymmetric, asymptotically flat exact solutions to Einstein's vacuum field equations, which are classified by an integer δ and Kerr metric is the solution of $\delta = 1$. The number of ring singularity on the equatorial plane is δ. The odd δ metrices contain the surfaces of event horizon but the even δ metrices do not. Except the Kerr metric, however, the space-time becomes singular at the poles on these surfaces.

References

Tomimatsu, A. and Sato, H.: 1972 *Phys. Rev. Letters* **29**, 1344.
Tomimatsu, A. and Sato, H.: 1973 *Prog. Theor. Phys.* **50**, 95.
Tomimatsu, A. and Sato, H.: *Lettere Nuovo Cimento*, to be published.

A NEW SOLUTION OF THE EINSTEIN-MAXWELL EQUATIONS FOR A SYSTEM WITH MASS, MAGNETIC MOMENT, CHARGE, AND ANGULAR MOMENTUM*

LOUIS WITTEN

University of Cincinnati, Cincinnati, Ohio, U.S.A.

Abstract. A five parameter solution of the combined Einstein-Maxwell equations is given which describes a source containing mass, electric charge, magnetic dipole, higher multipole moments of all three kinds, and angular momentum. The solution is asymptotically flat and has a singular infinite red shift surface. Possible relevance of the solution to black hole physics is discussed.

---

* Based on a paper entitled 'A Five Parameter Exterior Solution of the Einstein-Maxwell Field Equations' by F. Paul Esposito and Louis Witten, *Phys. Rev.* **D8**, 3302 (1974).

ACCRETION OF MATTER ONTO BLACK HOLES

R. A. SUNYAEV

Institute of Applied Mathematics, Moscow, U.S.S.R.

(Presented by M. M. Basko)

The theory of disk accretion is presented in the paper by Shakura and Sunyaev (1973). The structure of the disk, its luminosity and the spectrum produced are discussed. The reported quasiperiodic variability in the radiation from black holes is analysed in the paper by Sunyaev (1972). Reflection of X-rays and the possibility of observing X-ray pulsars which do not have a beam impinging on the Earth was considered by Basko *et al.* (in press).

The beaming of radiation by the accretion onto magnetic neutron stars is reported by Gnedin and Sunyaev (1973) and Basko and Sunyaev (in press). The beamed optical radiation from neutron stars due to the gyroemission of heavy ions and its connection with the observed optical pulsations of HZ Her are discussed by Gnedin and Sunyaev (in press). They also propose to search for the X-ray line in the spectrum of X-ray pulsars which corresponds to the electron gyrofrequency. In the paper by Cherepashchuk and Sunyaev (in press), new arguments in favor of the mass Cyg X-1 exceeding 3 $M_\odot$ are presented. The properties of other X-ray binary systems are discussed.

References

Basko, M. M. and Sunyaev, R. A.: *Astron. Astrophys.*, in press.
Basko, M. M., Sunyaev, R. A., and Titarchuk, L. G.: *Astron. Astrophys.*, in press.
Cherepashchuk, A. M. and Sunyaev, R. A.: *Monthly Notices Roy. Astron. Soc.*, in press.
Gnedin, Yu. N. and Sunyaev, R. A.: 1973, *Astron. Astrophys* **25**, 233.
Gnedin, Yu. N. and Sunyaev, R. A.: *Astrophys. Space Sci.*, in press.
Shakura, A. I. and Sunyaev, R. A.: 1973, *Astron. Astrophys.* **24**, 337.
Sunyaev, R. A.: 1972, *Astron. Zh.* **49**, 1153.

ACCRETION ONTO RELATIVISTIC OBJECTS

M. J. REES*

Astronomy Centre, University of Sussex, Falmer, Brighton BN1 9QH, England

Abstract. The physics of spherically symmetrical accretion onto a compact object is briefly reviewed. Neither neutron stars nor stellar-mass black holes are likely to be readily detectable if they are isolated and accreting from the interstellar medium. Supermassive black holes in intergalactic space may however be detectable. The effects of accretion onto compact objects in binary systems are then discussed, with reference to the phenomena observed in variable X-ray sources.

1. Introduction

Early studies of accretion (Bondi, 1952; Mestel, 1954 and earlier references cited therein) were motivated by the hope that infalling interstellar matter might cause a substantial increase in the luminosity of ordinary stars. The effects turned out to be generally insignificant, for two reasons: (i) the gravitational potential GM/r at the surface of an ordinary star is only $\sim 10^{-6} c^2$, implying an energy release per unit mass which is very small compared to that available from nuclear reactions; and (ii) the low density of surrounding matter means that the accretion rate onto an isolated star is low.

Accretion onto neutron stars or black holes ($GM/r \gtrsim 0.1\ c^2$) is obviously likely to be more significant, because each gram of infalling matter may yield $\gtrsim 10^{20}$ erg of electromagnetic energy; and a search for the radiation thus emitted is surely one of the most promising ways of detecting such objects. The recent evidence for compact objects in close binary systems has been a stimulus for much further work on this subject, because one here expects more spectacular observable effects, since there should be not only a large energy release per gram, but also a high accretion rate owing to the presence of a close companion star.

In this Paper I shall first briefly discuss accretion onto isolated compact objects, and then consider some theoretical implications of the remarkable X-ray observations which Giacconi has described for us.

2. Accretion Onto Isolated Compact Objects

The simplest case of accretion is that discussed by Bondi (1952), Shvartsman (1971) and Shapiro (1973a) in which the gravitating object is at rest relative to a surrounding medium. A characteristic length which enters into the problem is the so-called 'accretion radius'

$$r_A \simeq \frac{GM}{c_\infty^2} \simeq 4.8 \times 10^{13} \frac{M}{M_\odot} \left(\frac{T_\infty}{10^4\ \text{K}}\right)^{-1} \left(\frac{2}{1+X}\right) \text{cm,} \tag{1}$$

* Present address: Institute of Astronomy, Madingley Road, Cambridge CB3 0HA.

where c_∞ and T_∞ denote the sound speed and temperature in the medium at distances $\gg r_A$ from the gravitating object. r_A is the distance at which the escape velocity is $\sim c_\infty$. It is assumed in (1) that the gas consists predominantly of hydrogen (with ambient proton density n_∞ cm^{-3}) and that X is the fractional ionization. If the gas can be treated as a fluid, then the general nature of the flow (where v is the inward velocity) is

$$r \gg r_A : \begin{cases} n \simeq n_\infty \\ v \propto r^{-2} \end{cases}$$

$$r \simeq r_A : v \simeq c_\infty$$

$$r \gg r_A : \begin{matrix} v \propto r^{-1/2} \\ n \propto r^{-3/2} \end{matrix} \quad \text{(free fall)}$$

The precise details of the flow depend on the effective equation of state for the gas (and thus on whether – for example – radiative cooling is important), but the above is essentially correct if the behaviour of the gas can be characterised by a ratio of specific heats $\gamma < \tfrac{5}{3}$. For $\gamma = \tfrac{5}{3}$, we have a special case for which the solution is not unique. This is because the flow is not supersonic for $r \ll r_A$ as it is when $\gamma < \tfrac{5}{3}$ (since in this case $T \propto r^{-1}$ so the internal and kinetic energies increase at the same rate), so the *inner* boundary condition is relevant. The accretion rate turns out to be approximately

$$\dot{M} \simeq 2 \times 10^{10} \left(\frac{M}{M_\odot}\right)^2 \left(\frac{T_\infty}{10^4 \text{K}}\right)^{-3} \left(\frac{2}{1+X}\right)^{3/2} n_\infty \text{ gm s}^{-1}, \tag{2}$$

the precise value being dependent on the equation of state.

The above results remain adequate if the object is moving at speed V relative to the medium, provided that this motion is subsonic. If $V \gtrsim c_\infty$, c_∞ should be replaced by $(c_\infty^2 + V^2)^{1/2}$ in the formula for $\dot{M}$. In this latter case, the object accretes all the material within a cylindrical column of radius GM/V^2 so the accretion rate $\propto V^{-3}$. The flow pattern near the object itself – which is, of course, the region where most of the energy is liberated – may still be almost spherically symmetrical if angular momentum is unimportant. As was pointed out by Salpeter (1964), this process would gradually decelerate a fast-moving object.

2.1. Neutron Star

Where the accreting object has a 'hard' surface, as does a neutron star, all the kinetic energy of infall must emerge in radiation. This radiation amounts to $\sim 10^{20}$ erg gm^{-1} for a neutron star. The stopping distance of protons impacting on the surface at speeds $\gtrsim c/3$ is larger than the mean free path of outgoing photons. Thus in general the energy emerges as thermal radiation. When $n_\infty \simeq 1$ cm^{-3} the expected temperatures are in the range 10^5–10^6 K and the power output in soft X-rays $\sim 10^{30}$ erg s^{-1} (Ostriker et al., 1970). (As was emphasised by Shvartsman (1971), the appropriate value of T_∞ to take in (2) is always $\gtrsim 10^4$ K, because photoionization would be important out to radii $\gg r_A$. The problem must therefore be treated self-consistently, T_∞ being itself a function of n_∞). Even though X-rays from an individual isolated neutron star in the interstellar medium would not be detectable unless its distance were $\lesssim 10$ pc, the *integrated* X-ray emission from neutron stars accreting interstellar gas may in fact make a significant contribution to the soft X-ray background, because there may be $\gtrsim 10^9$ defunct pulsars in the galactic disc. Accretion onto neutron stars is discussed further in Section 3 in connection with binary X-ray sources.

2.2. Stellar-mass black holes

When the central object is a black hole, however, we will only observe the energy radiated by the material *before* it is swallowed by the hole. If the infalling matter has no angular momentum, this radiated energy essentially derives from the '$p\,dV$ work' done in compressing the gas. If the cooling time is *long* compared with the characteristic infall timescale, then Bondi's $\gamma = \frac{5}{3}$ solution is applicable, and only a small fraction of the energy will be radiated. For given conditions at infinity, this situation will prevail if the mass is *below* a certain critical value. Taking only bremsstrahlung and line cooling into account, this limit is given by

$$\frac{M}{M_\odot} \lesssim 10^4 \left(\frac{T_\infty}{10^4}\right)^{1/2} n_\infty^{-1}. \tag{3}$$

This is well satisfied for a stellar-mass black hole in the interstellar medium. Material accreted by such an object would therefore heat up, rendering the environment of the black hole a *very weak* source of X-rays and γ-rays. Shvartsman (1971) and Shapiro (1973a) have discussed this situation in more detail. An additional effect which reduces the intensity of the emergent radiation, calculated fully by Shapiro, arises from the fact that the radiation is generated very close to the black hole by material almost freely falling inwards. Both the gravitational redshift and aberration are then important. In later unpublished work Shapiro shows that accretion of matter with zero net angular momentum onto a Kerr black hole differs only slightly from the Schwarzschild case. At the high temperatures attained close to the black hole ($T \simeq 10^{12}$ K) the particle mean free paths are so long that a fluid-dynamical treatment is not really self-consistent unless collective effects are operative. However the effects of an interstellar magnetic field have to be taken into account. If this field is dynamically important at r_A, it obviously affects the character of the flow. But even if it is initially *negligible*, the field lines will be stretched radially during the inflow (instead of being compressed isotropically). As a consequence, the magnetic energy density varies as r^{-4}, and is likely eventually to become dynamically important (bearing in mind that the accretion radius is typically $\sim 10^8$ times larger than the Schwarzschild radius). The effects of this field are: (a) it ensures fluid-like behaviour of the material, because the gyroradius becomes very small compared with the scale of the system; (b) synchrotron radiation by electrons becomes an important cooling process (indeed, Shvartsman (1971) and Shapiro (1973b) estimate that about 1% of the energy of the infall will be radiated, yielding a luminosity $3 \times 10^{30} (M/M_\odot)^{3/2} n_\infty^{1/2}$ erg s^{-1} with a flat spectrum that cuts off in the visible, or possibly the ultraviolet band; (c) the magnetic field may induce turbulent motions, leading to more efficient dissipation of the infall energy. Although this is the simplest accretion problem of physical interest, several interesting aspects of it still await a thorough discussion. For example, electron and proton temperatures would tend to become unequal (the electrons become relativistic at $r \simeq 10^3 \; GM/c^2$, and so have an effective γ of $\frac{4}{3}$ – and are also subject to radiative losses – whereas the protons, being always non-relativistic, have a γ of $\frac{5}{3}$). It is unclear to what extent energy exchange between the species would equilibrate the temperature. (If this does happen, the effective γ is $\frac{13}{9}$ if radiative losses are negligible).

When the cooling time is very *short* compared with the infall timescale, one might imagine that more radiation would be emitted. But this is not necessarily so, the reason being that once the gas has cooled down to a temperature such that $kT \ll GM/r$, the '$p \, dV$ work' becomes a negligible fraction of the gravitational energy released. (The situation might, however, be changed if a small-scale magnetic field could maintain supersonic turbulent motions which dissipated via shockwaves; but this is another problem that has not yet been properly treated). The simple analysis suggests that the maximum radiative efficiency is attained in situations where the infall timescale and cooling timescale become comparable close to the Schwarzschild radius. Even then, mainly because of aberration and the gravitational redshift, the efficiency is only $\sim 1\%$.

Spherically-symmetric accretion onto an isolated collapsed object in interstellar space is thus unlikely to lead to any readily detectable radiation. (If a source of this kind *were* to be detected, it might be characterised by irregular rapid variability). This conclusion holds whether the gravitational field around the object is given by the Schwarzschild metric or by the more general Kerr metric. If the black hole moves at speed V, material would be captured from within a cylinder of radius $r_A (c_\infty/V)^{1/2}$ (for $V \lesssim c_\infty$). Interstellar material would have an initial angular velocity of perhaps $\sim 10^{-7}$ rad yr^{-1} on the scale of a few parsecs, and it is easily confirmed that provided $V \gtrsim 1$ km s^{-1}, the associated angular momentum remains dynamically unimportant compared to the gravity of the black hole even when the material approaches the Schwarzschild radius. As Shvartsman (1971) and Novikov and Thorne (1973) point out, this conclusion might alter if there were interstellar turbulence on scales down to $\sim 10^{14}$ cm. (Note that if $V=0$ we have a singular case when there is no strict steady state solution to the accretion problem, and angular momentum eventually becomes important for an arbitrarily small rotation velocity 'at infinity').

2.3. SUPERMASSIVE BLACK HOLES

Accretion onto isolated *supermassive* black holes was first discussed by Salpeter (1964). According to (2), the accretion rate goes as M^2, and this suggests that one might expect high luminosities from very massive collapsed objects even when the surrounding gas has the low density and high temperatures appropriate to the intergalactic medium. This topic was recently considered by Pringle et al. (1973), who showed that a galactic-mass collapsed object in a cluster of galaxies might have a luminosity of as much as $\sim 10^{46}$ erg s^{-1}. The reason for expecting this high luminosity lies partly in the high mass, but also results from the likelihood that the infalling matter would have so much angular momentum that it would not fall radially inward but would instead form a disc, from which a radiative efficiency in the range 6–42% seems guaranteed (see Section 3). The spectral distribution of the radiation is uncertain, and depends on whether the magnetic fields become high enough for synchrotron emission to be the major cooling process.

Lynden-Bell (1969) and Lynden-Bell and Rees (1971) have considered the observable effects of black holes of masses $\sim 10^8$ $M_\odot$ at the centres of galaxies. Such objects may be expected to form as the endpoint of the evolution of a quasar. Many of the observed forms of activity in galactic nuclei can be interpreted as effects of accretion of interstellar matter by a central black hole, but there are no strong reasons for favouring this model over the many others proposed to explain these phenomena.

3. Compact Objects in Binary Systems: General Considerations

I shall now attempt to review some of the theoretical implications of the remarkable new results which Giacconi has described in his paper (this volume, p. 147). Before doing so, however, it is perhaps worth recalling that the idea of X-ray sources being associated with close binary systems dates back to the earliest days of X-ray astronomy. It was Hayakawa and Matsouko (1964) and Zel'dovich and Guseynov (1965) who first made the suggestion that binary stars might be X-ray sources. After Sco X1 had been identified with an object reminiscent of an old nova, many other theorists proposed that X-ray sources involved transfer of matter from one star onto a compact companion (see Burbidge (1972) for an account of these developments). It is still unclear whether this is actually happening in Sco X1; but there now seems little doubt that it *is* the case for a major class of X-ray sources in the Galaxy.

What, then, can be said concerning the general nature of objects like Her X1, Cen X3, Cyg X1 and Cyg X3? The first general point is that the rapid variability suggests, though of course it does not prove, that a very small object – probably even smaller than a white dwarf – is involved. The gravitational potential well associated with such an object is very deep indeed, and accretion therefore provides an efficient energy source. If, as seems to be the case in the observed X-ray binaries, the compact object is in a close orbit – almost a grazing orbit – around another star, then a copious supply of material is available from the companion. Because the efficiencies are so high, the accretion rates need only be in the range 10^{16}–10^{18} gm s^{-1} (10^{-10}–10^{-8} $M_\odot$ yr^{-1}) in order to produce the observed luminosities. These are modest compared to the inferred transfer rates in other binary systems, and could be supplied by a stellar wind even if the companion star does not overflow its Roche surface.

A second general point is that most of the gravitational energy is liberated deep in the potential well – at or near the surface of the compact object if it is a neutron star; within a few Schwarzschild radii if it is a black hole. Thus the effective dimensions of the source (assuming that the compact object is in the stellar mass range) are only $\sim 10^6$ cm. If 10^{36}–10^{38} erg s^{-1} are radiated thermally from such a small region, a temperature high enough that the energy emerges predominantly in the X-ray band is therefore guaranteed.

In this report, I shall confine attention exclusively to the 'standard model' in which the X-ray source is regarded as being associated with either a neutron star or a black hole. Although, as Giacconi has described for us, the evidence favouring this model seems fairly compelling, the case is certainly not completely watertight; and one should bear in mind that some quite different interpretations for various aspects of these phenomena still remain tenable. I shall first outline the main features of the 'standard model' and then comment on how it may apply to some particular sources.

My reason for concentrating on this model is that it seems more plausible than any specific alternative so far proposed. Also this model has formed the basis for many detailed studies of these phenomena – indeed it will only be possible for me to give a sketch of most of this work, and some aspects of the problem will be left out entirely.

The X-rays binaries obviously involve all the problems connected with ordinary close binary systems – problems which are still ill-understood despite having been with us for many years – together with a whole range of new phenomena connected with the compact object. For convenience of exposition, it is convenient to split the subject into three parts: the mass transfer (relevant length scales $\sim 10^{11}$ cm); the accretion disc (dimensions $\sim 10^{10}$ cm); and the compact object itself, which is also the region where the X-rays are presumed to originate (10^6–10^8 cm). Unfortunately these three areas cannot be regarded as entirely disjoint, despite the very different length scales involved. For example, the X-ray intensity and spectrum is probably determined by processes occurring close to the compact object, but it may nevertheless have an important influence on the flow of matter from the companion because of heating and radiation pressure effects.

3.1. THE MASS TRANSFER ($\sim 10^{11}$ cm)

Much theoretical work has been based on the hypothesis that the companion star fills its Roche lobe, and that material flows across the Lagrangian point. It is important, however, to remember that these analyses are only strictly valid if the star corotates with the orbital period. This is probably quite a good assumption in these close systems, unless they were perturbed $\lesssim 10^6$ yr ago. Some calculations – for example, estimated limits on the masses of the X-ray sources – depend rather heavily on this postulate. It is also possible that the star does *not* fill its Roche lobe, but has a strong stellar wind. Gas streams may cause the optical emission lines observed in these systems (and also, incidentally, confuse attempts at radial velocity determinations). I shall say nothing about this topic, which is similar to that which arises in all close binary systems.

3.2. THE ACCRETION DISC ($\lesssim 10^{10}$ cm)

By whatever process material is captured from the companion star, it is likely to have so much angular momentum that it cannot fall directly onto the compact object. The matter will instead dissipate its motions perpendicular to the plane of symmetry and form a differentially rotating disc, the rotational velocity at each point being approximately Keplerian, and then gradually spiral inwards as viscosity transports its angular momentum outwards. If the companion star is overflowing its Roche lobe, it is conventionally assumed that the matter joins the disc at the radius where its angular momentum relative to the compact object equals that of a Keplerian orbit. This argument suggests that the so-called 'hot spot' appears at a radius which is $\sim 20\%$ that of the Roche lobe around the compact star. The structure of the outer part of the disc is not well understood. The disc must extend further out than the hot spot, because *some* of the material transferred from the companion star has to carry away the angular momentum – it cannot all be accreted by the compact object. A further complication is that the gravitational field of the companion star probably cannot be ignored in the outermost part of the disc, so the gas will not circulate in simple Keplerian orbits.

If the accreted matter is captured from a strong stellar wind, it will tend to have less net angular momentum; but the disc would still extend out to a radius $\gtrsim 10^9$ cm in general.

The structure of accretion discs has been discussed by many authors – for example Prendergast and Burbidge, 1968 (who considered a disc surrounding a white dwarf); Lynden-Bell, 1969; Pringle and Rees, 1972; Shakura and Sunyaev, 1973; Novikov and Thorne, 1973 – and the details will not be repeated here.

An obvious prerequisite for the existence of a disc (whose thickness must, by definition, be only a small fraction of its radius) is that radiative cooling should be efficient enough to remove most of the energy liberated by viscous friction, so that the internal energy is small compared with the gravitational binding energy, i.e.

$$kT\left(1+\frac{p_r}{p_g}\right) \ll \frac{GMm_p}{r}, \qquad (4)$$

where p_r/p_g is the ratio of radiation pressure to gas pressure, and m_p is the proton mass. For accretion flows with the parameters appropriate to X-ray sources the densities are high enough, and the timescales long enough, to ensure that (4) is almost certainly fulfilled. Also, the mass in the disc is gravitationally negligible compared to that of the central object.

If a steady state has been set up, the structure of the disc is governed by the following system of equations. First, the same mass flux $\dot{M}$ must flow across any radius r, so that

$$\dot{M} = 2\pi r \int \varrho(r,z)\, v_r(z)\, \mathrm{d}z \qquad (5)$$

for all r, when z is the coordinate perpendicular to the disc measured from the plane of symmetry.

A second, and somewhat less trivial, requirement is that the *flux of angular momen-*

tum should be the same at all r. Angular momentum is transported inward by the accreted matter, but transported outward by the viscous stresses. The difference between these quantities represents the rate at which the central compact object is gaining angular momentum. Following Novikov and Thorne (1973) we assume that angular momentum is being accreted at a rate $\beta \dot{M}(GMr_1)^{1/2}$, where r_1 is the radius of the inner boundary of the disc. Since the specific angular momentum deposited on the compact object cannot exceed the Keplerian value at r_1, we have $\beta \leqslant 1$. One then finds that the heat dissipated per unit surface area of the disc at a radius $r > r_1$ is

$$p(r) = \frac{3\dot{M}}{4\pi r^2} G \frac{M}{r} \left(1 - \beta \left(\frac{r_1}{r}\right)^{1/2}\right). \tag{6}$$

It is important to note that β is a second parameter which is not completely determined by $\dot{M}$ – one can imagine situations with the same $\dot{M}$ but different torques in the disc, and therefore different values of $p(r)$. When $r \gg r_1$, however, one finds, independently of β, that the energy radiated at radii $\geqslant r$ is 3 *times larger* than the energy lost by the accreted material while spiralling inward to that radius. The extra contribution arises because the viscous stresses transport *energy* outward as well as momentum. One might at first sight worry about the energy budget for the disc as a whole. However, when $\beta = 1$ one finds that the total energy radiated, integrating over all $r \geqslant r_1$, is precisely equal to M multiplied by the binding energy of Keplerian orbit of radius r_1; when $\beta = 0$, the factor of 3 enhancement applies right in to $r = r_1$, but in this case the extra energy comes from viscous torques which apply a drag to the compact object.*
(i.e. twice as much energy in this case is supplied by the central spinning object as comes from the infalling material itself.)

These deductions do not depend on the viscosity – if this is low, then the radial velocity v_r is small, so the equilibrium value of ϱ needed in order to give a given $\dot{M}$ must be high; and conversely. But to analyse the structure of the disc in any further detail one *must* know something about the viscosity, and this is the stumbling-block to further progress. Possible causes of viscosity include turbulence induced by the differential rotation, convective motions, or sheared magnetic fields. Pringle and Rees (1972) and Shakura and Sunyaev (1973) made specific simplifying assumptions about the viscosity, which enabled them to discuss the vertical structure of the disc (i.e. the balance between the pressure gradient perpendicular to the disc and the component of gravity in that direction), and the spectrum of the emergent radiation. However one has little confidence that one knows even the appropriate order of magnitude for the viscosity, and it therefore seems premature to discuss the spectrum of the disc in great detail. The dominant emission mechanism is probably thermal bremsstrahlung, though the spectrum may be appreciably distorted as a result of scatterings by the hot thermal electrons. All that can be said is that the effective tem-

---

* Note that the above analysis is strictly Newtonian. When one considers an accretion disc surrounding a black hole, then one finds that the energy radiated by the disc *equals* the energy lost by infalling matter when the black hole accretes a specific angular momentum appropriate to the circular orbits at the inner edge of the disc (see Novikov and Thorne (1973) for the details of the relativistic case).

perature must be at least as high as the black body temperature needed to radiate a power $p(r)$.

3.3. THE COMPACT OBJECT AND THE X-RAY EMISSION (10^6–10^8 cm)

When the compact object is a black hole, the disc extends inward to the innermost stable circular orbit, the emission being concentrated within a few Schwarzschild radii. As discussed by Bardeen (this volume, p. 132) efficiencies of up to 40% are possible for accretion discs around Kerr black holes, the precise upper limit depending on how much of the emitted radiation is captured by the hole. Attempts to determine the expected radiation spectrum are impeded by our ignorance about the viscosity, which introduces far larger uncertainties than those corresponding to the difference between a Schwarzschild and extreme Kerr black hole. In general, the temperature decreases outwards and, even though the emission is thermal, the integrated spectrum may resemble a power law. Radiation from the outer parts of the disc would not be energetically significant unless, as discussed by Shakura and Sunyaev (1973) the disc is so thick in relation to its radius that X-rays from the inner regions are intercepted by the disc and reradiated at softer energies. Some further aspects of this model, as it may apply specifically to Cyg X1, are discussed later.

When the central object is a spinning, magnetised neutron star, a far more complex situation ensues, which has been discussed extensively by Pringle and Rees (1972), Davidson and Ostriker (1973), and Lamb *et al.* (1973). If the neutron star were unmagnetised, then the disc would extend inwards until the accreted material grazed the star's surface. If, however, the neutron star has a surface magnetic field of the same strength as is inferred for pulsars ($\sim 10^{12}$ G) then the magnetic stresses will influence the dynamics out far beyond the surface of the star. We define the 'Alfvén radius' to be that distance at which the magnetic stresses are comparable with the viscous stresses in the disc, i.e.

$$\frac{(H(r_A))^2}{4\pi} \simeq \varrho(r_A)\, v_r(r_A)\, v_\theta(r_A).$$

The Alfvén radius depends on $\dot{M}$, but somewhat insensitively because H^2 depends on r at least as steeply as r^{-6}; and for typical parameters r_A is 10–100 times larger than r_*. r_A is fortunately independent of the viscosity except insofar as this affects the scale height. This means that the radiation from the disc itself is relatively unimportant. Once matter penetrates within r_A, it is constrained to follow the field lines. If the star has an oblique dipole field, the infalling matter will impact on the surface in the vicinity of the magnetic polar caps. The physics at $r \simeq r_A$ is so complicated that one cannot really estimate which of the magnetic field lines can capture matter. These field lines will probably, however, be a subset of those which would have reached out to radii $\gtrsim r_A$ in the absence of infalling plasma. This suggests that, when $r_A \gg r_*$, the material will be channelled onto only a small fraction of the stellar surface.

The dominant radiation mechanisms would be bremsstrahlung or cyclotron radiation (including emission at the first few harmonics of the basic cyclotron frequency).

Lamb *et al.* (1973) and Gnedin and Sunyaev (1973) have discussed the likely beam shape of the emergent radiation. If the dominant opacity were ordinary Thomson scattering, the radiation would tend to leak out of the sides of the accretion column, yielding a fan beam. If the magnetic field is so strong that the cyclotron frequency exceeds the radiation frequency under consideration, then electron scattering is inhibited for radiation propagating along the field direction, and also for radiation travelling across to the field which is polarised such that the electric wave vector is at right angles to the magnetic field. Realistic models can yield either pencil beams or fan beams, depending on the strength of the magnetic field and the polarization of the radiation. Modulation of this beam pattern each time the neutron star spins generates the X-ray pulse shape. The radiation would generally be expected to display a high degree of both linear and circular polarization.

Some other aspects of this scheme are discussed later in connection with particular sources.

3.4. The 'Eddington Limit'

An important role in these models is played by the so-called 'critical luminosity' or 'Eddington limit' at which radiation pressure balances gravity. If Thomson scattering provides the main opacity and the relevant material is fully ionized, then this luminosity is

$$L_{edd} = \frac{4\pi G M m_p}{\sigma_T c} \simeq 10^{38} \left(\frac{M}{M_\odot}\right) \text{erg s}^{-1}, \tag{7}$$

σ_T being the Thomson cross-section.

One might therefore expect that the accretion rate $\dot{M}$ could approach, but in no circumstances exceed, the value needed to yield this luminosity. Recently, Margon and Ostriker (1973) have in fact analysed the data on X-ray sources, and find that there does indeed seem to be a luminosity cut-off at around the expected value of L_{edd} for $M \simeq M_\odot$, and that there is a class of sources whose luminosities cluster close to this value. Because this issue is an important one, it is perhaps worth pointing out that the 'Eddington limit' is physically significant only under relatively restrictive circumstances – circumstances which are *not* generally met by the kinds of X-ray source models usually considered.

As emphasized by Buff and McCray (1974) the luminosity of a source powered by accretion cannot even approach L_{edd} if the effective cross section per electron is larger than σ_T. This is quite likely to be the case for a source emitting soft X-rays, because the relevant opacity (unless all the ions are completely stripped) is then primarily due to photoionization, for which $\sigma \gg \sigma_T$. If the value of $\dot{M}$ in binary X-ray sources is controlled by processes occurring near the surface of the companion star or the critical Roche surface, as in the 'self-excited wind' hypothesis (Basko and Sunyaev, 1973; Arons, 1973) then one might expect the luminosity to stabilise at a value *well below* L_{edd}.

There are, however, several types of situation where luminosities $\gg L_{edd}$ are possible especially under the extreme conditions prevailing near compact objects. Some of these are mentioned below:

(i) The effective opacity may be much *less* than that provided by Thomson scattering. In the context of X-ray sources this may, for instance, happen in the accretion column above the magnetic polar caps of neutron stars, where the scattering cross section is $\ll \sigma_T$ for photons below the cyclotron frequency travelling along the magnetic field direction.

(ii) The Eddington limit can also be violated in any non-spherically-symmetric configuration. Consider again, for example, the accretion column near a magnetised neutron star. If the magnetic field does *not* modify the opacity and make the scattering highly anisotropic, then radiation will tend to escape from the *sides* of the column. This means that the radiation flux along the column, and therefore the pressure opposing gravity, is then less than it would be in an isotropic situation. (An analogous argument may also apply to accretion discs).

(iii) As has been pointed out by Lamb et al. (1973) there are conceivable circumstances when the luminosity may exceed L_{edd} even when the appropriate cross section is σ_T *and* the accretion is isotropic. This is because $L > L_{edd}$ is merely the condition that infalling matter should be *decelerated*. But unless the total optical depth is sufficiently large, this does not guarantee that radiation pressure can halt the accretion. The infalling matter carries momentum across a sphere of radius r at a rate $\dot{M}v(r)$, where $v(r)$ is of the order of the free fall speed. If its kinetic energy is converted into radiation at a radius r_{min} the outward momentum flux, ignoring relativistic corrections, is $\sim (\dot{M}/2c)(v(r_{min}))^2$ (and less, of course, if the conversion efficiency is low). This means that the average photon must undergo more than $2c/v(r_{min})$ scatterings if radiation pressure is to stem the accretion flow (unless the main contribution to the opacity comes from radii $r \gg r_{min}$).

(iv) The Eddington limit is of course irrelevant in an *unsteady* or *explosive* situation: it is, for instance, violated by factors $\sim 10^5$ in supernovae.

A luminosity exceeding L_{edd} obviously entails a correspondingly higher accretion rate. If, however, the accretion has the high efficiency expected in X-ray sources, the observations seem to rule out a value of $\dot{M}$ of $\gtrsim 10^{-7}$ yr^{-1} in all cases. If the companion star loses mass at a higher rate than this (as is likely at certain stages of stellar evolution, and as would seem required if the changes of *orbital* period in Cen X3 are attributable to mass loss from the companion star) then most of the material would presumably escape from the system.*

4. Phenomena Observed in Particular Binary X-Ray Sources

Her X1 and Cen X3 are clear candidates for systems when the X-ray source is a

---

* The possibility has been raised (Zel'dovich et al., 1972; Ruffini and Wilson, 1973) that a neutron star may be able to accrete at a rate $\sim 10^{-3}$ $M_\odot$ yr^{-1} and get so hot that the energy escapes mainly as neutrinos, but the photon luminosity remains below L_{edd}. There seem severe doubts, however, about whether this situation could actually be set up by gradually increasing M, and whether it would be stable.

neutron star (and it is gratifying that the mass of Her X1 seems to be within the allowable range for neutron stars). There are several observed phenomena which can be tentatively explained on the basis of this model. These systems are like pulsars in that rotation provides the 'clock'. However the X-ray power radiated derives *not* from rotational kinetic energy – which could maintain the observed X-ray luminosity of Her X1 for $\lesssim 10$ yr – but from accretion. (This, as Schwartsman has pointed out, suggests at least part of the reason why pulsars are not found in binary systems. An isolated spinning neutron star, surrounded only by diffuse interstellar gas, generates the electromagnetically driven relativistic wind which is believed to be a precondition for the coherent pulsed radio emission. When such an object is embedded in a denser environment, the pressure of the relativistic outflow cannot hold the external matter at bay, and we instead get accretion, manifesting itself in the emission of thermal X-rays. One can estimate that Her X1 would have displayed pulsar-like behaviour only if its period were $\lesssim 0.1$ s).

4.1. Changes in the Pulse Period

Since an accreting neutron star is not drawing on its rotational energy as its main power supply, it is not obvious whether its spin rate should slow down or speed up. An element of gas accreted by the star carries angular momentum corresponding to corotation at the Alfvén radius. This suggests that the spin rate would speed up on a timescale

$$\frac{M}{\dot{M}}\left(\frac{r}{r_A}\right)^2. \qquad (8)$$

Even though $M/\dot{M} \simeq 10^8$ yr, this 'lever-arm' effect certainly allows a speedup as rapid as that observed in Cen X3. There is, however, a possible opposing effect tending to *brake* the rotation: this is the viscous torque exerted by the accretion disc outside r_A. These two effects can be of the same order of magnitude if

$$\left(\frac{GM}{r_A}\right)^{1/2} \simeq \Omega r_A$$

(and of course if $(2GM/r_A)^{1/2} < \Omega r_A$ it would be energetically possible for material at the inner edge of the disc to be flung out of the system by magnetic forces, leading to a further braking effect). Davidson and Ostriker (1973) suggest that Ω tends asymptotically to a value such that the net torque on the neutron star is zero. This value of Ω depends on r_A, which is itself a function of $\dot{M}$. Therefore, if there were fluctuations in the accretion rate, then Ω would tend to increase (decrease) as $\dot{M}$ increases (decreases). If Her X1 were close to this equilibrium state, and the fluctuations in $\dot{M}$ were small in amplitude, one could perhaps understand why Ω has been observed both to increase and to decrease, and why the timescale for these changes is slower than is the case for Cen X3.

The effects mentioned above are the dominant ones for causing changes in period.

Other effects – for example, the spin-up due to the contraction of the star as it accretes mass – occur on the much slower timescale of $M/\dot{M}$.

4.2. Her X1: the long-term variability

Giacconi has summarized the X-ray data on the '35 day cycle' and I shall here briefly mention some of the numerous suggestions already made to explain this puzzling behaviour. In this connection, it is important to recall that optical observations impose an important constraint on such suggestions. It appears that the 1.7 day period light variations persist throughout the 35 day cycle with more or less the same amplitude (even though a 35 day periodicity may be discernable in some of the fine details of the light curve (Kurochkin, 1973; Boynton et al., 1973)). Since the thermal inertia of the relevant layers of the companion star is small, this implies that the heating mechanism operates throughout the ~ 23 days out of ~ 35 when UHURU detects no X-rays from Her X1.

4.2.1. *Modulations in Mass Transfer Rate*

One class of theory for the 35 day cycle involves supposing that the *mass transfer* is modulated with this period. It seems unlikely that this could be due to some pulsation of the companion star because the expected pulsation periods would be $\ll 35$ days. Another possibility (Pringle, 1973; Henriksen et al., 1973) is that the spin period of the companion differs by $\sim 5\%$ from the orbital period. If the star displayed some departures from axisymmetry – a 'magnetic spot' associated with an especially vigorous wind for instance – then the transfer rate could vary with a synodic period of 35 days.

Conceivably some kind of feedback process may be operating. McCray (1973) has developed an ingenious model which utilises the fact that the X-ray luminosity is a significant fraction (perhaps $\sim 10\%$) of L_{edd}. When the X-rays are 'on', the X-ray source behaves with respect to the surrounding gas as though it had a somewhat lower mass. The 'effective' Roche lobe around the companion star might then expand so that material no longer overflowed it. Mass transfer would then cease, and no material would be added to the disc. The disc would then drain away, and the X-ray emission would stop. Mass transfer would then begin again, the disc would be replenished, and so on. McCray speculates that some kind of limit cycle is set up. The time-scale of this cycle would be determined by the length of time taken for a typical element of gas to spiral inward to the central object. A period of the general order of 35 days would certainly not be unreasonable, but one cannot claim to 'predict' it, because of the wide uncertainty about the efficiency of viscosity in the disc.

A fully developed theory along these lines must also take account of a competing process which might cause *positive* feedback. This arises because the X-rays, by heating the surface layers of the companion star, tend to *raise* the mass transfer rate by increasing the scale height in the atmosphere and/or by stimulating an enhanced stellar wind (Arons, 1973; Basko and Sunyaev, 1973; Alme, 1973). It has in fact been proposed (Lin, 1973) that the 35 day cycle could result from this type of positive feedback if the X-rays stimulate a mass transfer rate which 'overshoots' to such an extent that opacity effects around the compact object quench the X-rays. Plainly a proper theory of the 35 day cycle along the above lines must await a fuller understanding of how the X-rays interact with the companion star, and also of the factors that determine the residence time of material in the accretion disc.

4.2.2. Processes Occurring in the Accretion Disc

Katz (1973) has suggested that the rim of the accretion disc may not lie in the orbital plane of the system. This might happen if the companion star possessed a component of spin angular momentum which was not aligned with the orbital angular momentum. In this situation, the rim of the disc would precess, and could obscure the X-rays for some fraction of each precession period. To obtain a precession period of 35 days, Katz has to assume that the disc extends outwards to a larger radius than is customarily supposed.

It is also conceivable that the disc might be subject to convective or other instabilities which might cause it to dump material periodically onto the central object.

4.2.3. Modulation of Inflow from Alfvén Radius

Pines *et al.* (1973) have developed a model according to which the neutron star undergoes free precession in such a way that the angle between the magnetic axis and the plane of the accretion disc varies periodically. When this angle is small, accretion along the 'magnetic funnel' can proceed; but when the magnetic axis points too far out of the plane of the disc accretion is suppressed, and material transferred from the companion accumulates in the disc outside the Alfvén radius. It is not clear how large the precession amplitude would have to be in order for such an 'accretion gate' to operate. However Pines *et al.* list some other reasons why the accretion flow near the Alfvén surface could be sensitive to the orientation of the neutron star's rotation axis, so it is conceivable that a wobble through only a few degrees could suffice. On the basis of this model, Pines *et al.* have attempted to explain the other features of the 35 day cycle. The asymmetry between the sharp rise and the gradual fall in X-ray intensity during the 12 day 'on' period is readily explained. Matter accumulating during the 'off' period will be opaque to the X-rays until it has been photoionized. The fact that the switch-off occurs near orbital phases 0.25 or 0.75 is attributed to the higher density of obscuring matter along the line joining the two stars, which makes it more likely that the first X-ray to be seen will escape perpendicular to this line. The hypothetical 'hot spot' where the gas stream merges with the disc may be thick enough to obscure the X-rays at the phase of the orbit when it lies along our line of sight. The outer radius of the disc would decrease during the 'on' period, and the location of the hot spot would change (it is claimed) in such a way that the dip 'marches' in phase in the matter observed. The apparent tendency of the small amplitude 1.24 s *optical* pulsations (which probably come from gas with cooling time $\leqslant 1.24$ s which is being heated by the X-rays) to occur at particular particular orbital phases can also be explained.

4.2.4. Precession of Pencil Beam

Another idea involving precession of the neutron star (Brecher, 1972; Strittmatter *et al.*, 1973) is that the X-rays remain 'on' for the whole 35 day cycle, but that they emerge in a pencil beam which sweeps through our line of sight only for 12 days out of 35. There are some geometrical difficulties associated with this idea. In particular, the broad and relatively smooth observed X-ray pulse profile tells us something about the shape of the beam, and it is hard to reconcile this with the sharp onset of the high state or with the apparent lack of any marked systematic changes in the pulse shape during the 'on' state. A very large wobble amplitude ($\gtrsim 45$ deg) would certainly seem required by this model.

At least in models (a) and (c), the continuous heating of the companion star can only be explained by invoking a steady heat source. One possibility (Avni *et al.*, 1973) is that the neutron star emits a steady non-pulsed flux of soft X-rays, powered by the ~ 8 MeV per nucleon resulting from nuclear fusion of the accreted matter. This energy is liberated well below the neutron star surface, and emerges isotropically. But a serious problem arises with any model in which soft ($\lesssim 0.5$ keV) X-rays play the dominant role in the heating, because these photons (unlike harder X-rays) are absorbed predominantly *above* the photosphere. The associated energy input would then distort the temperature stratification, resulting in the formation of strong emission lines and suppression of the ordinary stellar absorption spectrum (Basko and Sunyaev, 1973; Strittmatter, 1973). It seems more likely that the star HZ Her is heated mainly by *hard* ($\gtrsim 10$ keV) X-rays, though the problem then is the inefficiency resulting from the high albedo (unless one considers photons of $\gtrsim 0.5$ MeV). Heating by fast particles is another possibility. In models (b) and (d), one may suppose that X-rays always hit the companion star even when they cannot propagate along our line of sight (though this requirement places further constraints on the geometry). A more attractive variant of (d) might be to postulate that the star is heated by hard X-rays which are not so strongly beamed as those detected by UHURU. This is theoretically plausible because the circumstance which might most naturally cause a pencil beam – the reduced scattering cross section for photons travelling along the magnetic field direction – would not be so effective at high photon energies.

35 days is much too short a free precession period for a neutron star with a liquid core. However a neutron star with a *solid* core and the 1.24 s spin period appropriate to Her X1 could plausibly sustain a sufficient deviation from axisymmetry to yield a 35 day precession, and would then automatically be rigid enough to be able to wobble through a large angle. (Mechanisms for exciting this kind of wobble and for sustaining it against damping processes are discussed by Pines (1973).)

In assessing the various models for the 35 day cycle it is of course crucial to know just how regular a phenomenon it really is. It is also relevant that Cen X3 displays extended lows which are apparently *not* strictly periodic. Finally, some explanation is also required for the *very* long (~ 10 yr) timescale variability in Her X1 (Jones *et al.*, 1973). If one were optimistic, one might therefore hope that *two* of the possibilities mentioned above might actually be relevant!

4.3. ORBITAL PERIOD OF CEN X3

The changes in the orbital period of Cen X3 have been attributed to a mass loss or mass transfer rate far higher than the minimum needed to provide the X-ray power. However this large amount of gas, even if it were escaping from the system, would cause so much opacity that the X-ray source could not be observed. A more plausible possibility (Pringle, private communication) is that angular momentum is still being exchanged between the orbital motion and the spin of the companion star. This suggests that the event which formed the neutron star happened within the last $\sim 10^4$ yr, and that the timescale of the observed orbital period change corresponds to that required for tidal effects to establish synchronous rotation.

4.4. CYG X1

This is the prime candidate for being an X-ray source involving a black hole. One would expect the accretion disc around a black hole to be subject to various instabilities: thermal instabilities, magnetic instabilities (perhaps analogous to those which Parker has discussed in the context of the interstellar gas in our Galaxy), or perhaps instabilities resulting from irregularities in the mass transfer rate. These could give rise to irregular flickering on all time scales down to the orbital period associated with the most tightly bound stable circular orbits, but no regular period would be expected. Even if one had no evidence on its mass, one would therefore suspect Cyg X1 of being a disc around a black hole, and it is therefore gratifying that the evidence on its mass strongly supports this interpretation. Sunyaev (1973) proposed that attempts should be made to search for pulse trains due to regions of enhanced emissivity orbiting the hole. The typical orbital periods would be $\sim 0.5 \, (M/M_\odot)$ ms for a Schwarzschild black hole, but ~ 8 times faster if the black hole had a maximal Kerr metric whose angular momentum was aligned with the disc, but with the same mass. (Further interesting complications can occur if the black hole is obliquely oriented relative to the disc. Some of these have been mentioned by Bardeen in his review (this volume, p. 131)).

Further information would be derived if an X-ray spectral feature originating in the disc could be discovered and its profile measured, but this seems unlikely to be feasible before 1980. It is important to remember that black holes are a consequence of almost all 'viable' theories of gravity, and much further work is needed before one can diagnose whether the properties of a given black hole agree better with those expected on the basis of general relativity than with the predictions of a rival theory.

4.5. SCO X1

It is still unclear whether Sco X1 belongs to the same family as the other X-ray sources. Basko and Sunyaev have argued that Sco X1 could have a normal star as a binary companion with $L \lesssim L_\odot$. However, because the X-ray output is $\sim 10^4 \, L_\odot$, the geometry and orientation of the system must be carefully specified in order to reduce the X-ray heating of the hypothetical companion to an acceptably low level.

The general spectrum of Sco X1 (in the infrared and optical bands as well as in

X-rays) is very well fitted by an accretion disc model (Pringle and Rees, 1972). Maybe Sco X1 could be a compact object surrounded by a *massive* disc, the disc being a remnant of a companion star destroyed by tidal forces (c.f. Faulkner, 1971).

4.6. Cyg X3

The gradual and incomplete character of the X-ray eclipses suggests that in this system the eclipse is caused not by the surface of the companion star, but by scattering and absorption in a strong wind. The observed 2.2μ infrared variations imply that the relevant layer of the heated side of the companion star has a temperature $\gtrsim 10^6$ K. This, however, is quite possible if one is seeing emission from the wind, which is heated to this temperature (Pringle, 1974).

Although Cyg X3 has a shorter period than the other X-ray binaries, the period is longer than that of systems such as DQ Herc. It may differ from such systems merely in having a neutron star (or black hole) as the compact component, instead of this being a white dwarf.

5. Concluding Remarks

Many important and interesting aspects of binary X-ray sources have not even been touched on in the foregoing review.

There is a whole complex of problems associated with the *optical light curve* of these systems. Heating of the companion star causes the side facing the compact object to be hotter and brighter than the eclipsed side. A full understanding of this effect involves detailed computations (along the lines of those already done by Arons (1973) and Basko and Sunyaev (1973)) of the structure of a stellar atmosphere irradiated by X-rays. A second quite different effect which leads to optical variations with *half* the orbital period arises from the distortion of the companion star by the compact object's gravitational field. Interpretation of actual light curves is complicated by further effects (emission by gas streams, radiation and absorption by the accretion disc itself, etc.) and one suspects that detailed model-building may prove somewhat fruitless unless some very clear-cut correlations between X-ray and optical variability are found. In Her X1, X-ray heating (or heating by some other radiation flux emanating from the compact object) is the dominant effect, and the effects of gravitational distortion are relatively minor; in Cyg X1, where the companion star is much more luminous relative to the X-ray source than is the case for Her X1, the heating augments the stellar luminosity by only $\sim 2\%$. The occurrence of X-ray heating, however, sets a *lower limit* to the apparent brightness of the optical counterpart for any eclipsing X-ray source. If there were no interstellar absorption, any X-ray source with an intensity of C UHURU counts which is observed to eclipse for a fraction f of every period should have an optical apparent magnitude

$$m \lesssim 15 - 2.5 \left(\log\left(\frac{C}{10}\right) + 2 \log(4f) \right).$$

Thus any eclipsing source in the UHURU catalogue would be optically identifiable were it not for the often severe effects of interstellar extinction.

The radio properties of the X-ray binaries are completely unexplained. However the fact that – even in the extreme case of Cyg X3 at the peak of its radio flare – the radio luminosity is a tiny fraction of the X-ray output, suggests that to concern ourselves with the details of the radio variability may be as premature as it would be to worry about solar flares before understanding the basic elements of stellar structure. Moreover, some binary systems containing two relatively normal stars have similar radio properties and this suggests that the radio behaviour is unlikely to be intimately connected to the compact object itself.

The existence of these close binary systems with compact components raises many astrophysical questions. How do they fit into the general scheme of binary star evolution? How do they evolve to their present state (and, in particular, how did they avoid disrupting during the catastrophe which formed the collapsed component)? Why, nevertheless are there only ~ 30 such systems in the Galaxy? What will be their eventual fate? – for example, what happens if a neutron star accretes so much material that it comes to exceed the limiting mass; or what happens when, later in its evolution, the companion star swells up and engulfs the compact object?

There is a real possibility of discovering an isolated massive black hole (as discussed in Section 2) by observing the effects of accretion. But X-ray binaries are the sole *known* instances of accretion onto relativistic systems. Far more detailed information can be expected from the next generation of X-ray detectors (and from more refined optical observations) so there seems little doubt that these objects will remain at the focus of theoretical attention for several years to come, and will allow us to check the theory of black holes (and neutron stars) against observations in many key respects.

References

Alme, M.: 1973, paper presented at conference on *Physics and Astrophysics of Compact Objects*, Cambridge, England.
Arons, J.: 1973, *Astrophys. J.* **184**, 539.
Avni, Y., Bahcall, J. N., Josse, P. C., Bahcall, N. A., Lamb, F. K., Pethick, C. J., and Pines, D.: 1973, *Nature Phys. Sci.* **246**, 36.
Basko, M. M. and Sunyaev, R. A.: 1973, *Astrophys. Sp. Sci.* **23**, 117.
Bondi, H.: 1952, *Monthly Notices Roy. Astron. Soc.* **112**, 195.
Boynton, P. E., Canterna, R., Crosa, L., Deeter, J., and Gerend, D.: 1973, *Astrophys. J.* **186**, 617.
Brecher, K.: 1972, *Nature* **239**, 325.
Buff, J. and McCray, R. A.: 1973, *Astrophys. J.*, in press.
Burbidge, G. R.: 1972, *Comm. Astrophys. Space Phys.* **4**, 105.
Davidson, K. and Ostriker, J. P.: 1974, *Astrophys. J.* **179**, 585.
Faulkner, J.: 1971, *Astrophys. J. Letters* **170**, L99.
Gnedin, Y. N. and Sunyaev, R. A.: 1973, *Astron. Astrophys.* **25**, 233.
Hayakawa, S. and Matsouko, M.: 1964, *Prog. Theor. Phys. Suppl.* **30**, 204.
Henriksen, R. N., Reinhardt, M., and Aschenbach, B.: 1973, *Astron. Astrophys.* **28**, 47.
Jones, C. A., Forman, W., and Liller, W.: 1973, *Bull. Am. Astron. Soc.* **5**, 32.
Katz, J. I.: 1973, *Nature Phys. Sci.* **246**, 87.
Kurochkin, N. E.: 1973, *Inform. Bull. Var. Stars*, No. 55.
Lamb, F. K., Pethick, C. J. and Pines, D.: 1973, *Astrophys. J.* **184**, 271.

Lin, D. N. C.: 1973, *Astron. Astrophys.* **29**, 109.
Lynden-Bell, D.: 1969, *Nature* **223**, 690.
Lynden-Bell, D. and Rees, M. J.: 1971, *Month.y Notices Roy. Astron. Soc.* **152**, 461.
Margon, B. and Ostriker, J. P.: 1973, *Astrophys. J.* **186**, 91.
McCray, R. A.: 1973, *Nature Phys. Sci.* **243**, 94.
Mestel, L.: 1954, *Monthly Notices Roy. Astron. Soc.* **114**, 437.
Novikov, I. D. and Thorne, K. S.: 1973, in C. deWitt (ed.), *Black Holes*, Gordon & Breach, p. 343.
Ostriker, J. P., Rees, M. J., and Silk, J. I.: 1970, *Astrophys. Letters* **6**, 179.
Pines, D.: 1973, report presented at *16th Solvay Conference*.
Pines, D., Lamb, F. K., and Pethick, C. J.: 1973, *Proc. N. Y. Acad. Sci.*, in press.
Prendergast, K. H. and Burbidge, G. R.: 1968, *Astrophys. J. Letters* **151**, L83.
Pringle, J. E.: 1973, *Nature Phys. Sci.* **243**, 90.
Pringle, J. E.: 1974, *Nature* **247**, 21.
Pringle, J. E. and Rees, M. J.: 1972, *Astron. Astrophys.* **21**, 1.
Pringle, J. E., Rees, M. J., and Pacholczyk, A. G.: 1973, *Astron. Astrophys.*, in press.
Ruffini, R. and Wilson, J.: 1973, *Phys. Rev. Letters* **31**, 1362.
Salpeter, E. E.: 1964, *Astrophys. J.* **140**, 796.
Shakura, N. I. and Sunyaev, R. A.: 1973, *Astron. Astrophys.* **24**, 337.
Shapiro, S. L.: 1973a, *Astrophys. J.* **180**, 531.
Shapiro, S. L.: 1973b, *Astrophys. J.*, in press.
Shvartsman, V. F.: 1971, *Soviet Astron. AJ* **15**, 377.
Strittmatter, P. A.: 1973, *Astron. Astrophys.* **185**, 69.
Strittmatter, P. A., Scott, J., Whelan, J., Wickramasinghe, D. T., and Woolf, N. J.: 1973, *Astron. Astrophys.* **25**, 275.
Sunyaev, R. A.: 1973, *Soviet Astron. AJ* **16**, 941.
Zel'dovich, Ya. B. and Guseynov, O. K.: 1965, *Astrophys. J.* **144**, 840.
Zel'dovich, Ya. B., Ivanova, L. N., and Nadezhim, D. K.: 1972, *Soviet Astron. AJ* **16**, 209.

ON A POSSIBLE INFLUENCE OF MAGNETIC FIELDS ON THE STRUCTURE OF A DISK FORMED DURING ACCRETION OF PLASMA IN BINARY SYSTEMS

L. A. PUSTILNIK and V. F. SHVARTSMAN

*Special Astrophysical Observatory, U.S.S.R. Academy of Sciences,
St. Zelenchukskaya, Stavropolskij Kraj 357140, U.S.S.R.*

Abstract. During accretion of plasma in binary systems containing compact objects magnetic fields are most likely to become arranged quickly, grow and fully determine the disk structure. The disk divides into separate dense clots, and a corona appears over the equatorial plane of the system. In the corona magnetic lines of force reconnect and beams of relativistic particles are generated.

Reference

Pustilnik, L. A. and Shvartsman, V. F.: 1973, *Commun. SAO U.S.S.R. AS*, in press.

WHAT INFORMATION CAN BE EXTRACTED FROM RADIO DATA ABOUT THE EXISTENCE OF SUPERMASSIVE BLACK HOLES?

L. M. OZERNOY

P. N. Lebedev Physical Institute, Academy of Sciences of U.S.S.R., Moscow, U.S.S.R.

The most convincing arguments to prove or disprove the idea that a supermassive accreting black hole serves as an energy source for quasars and quasar-like phenomena in galactic nuclei may be extracted from the variability of their radiation. There are at present numerous data on optical and radio variability of a number of objects.

As for optical data their statistical analysis applied to the quasar 3C 273 showed that the source of activity is *not* a cluster of independently and accidentally flaring objects like supernovae, but is some coherent body (Gudzenko et al., 1968, 1971). At present there are about ten quasars and active galactic nuclei demonstrating the quasi-periodic character of their optical variations (they are listed in Ozernoy, 1973). Such a behavior of luminosity supports the conclusion that the source of activity is a single body.

The next problem to be solved is whether the source of activity is a collapsed body or not. The discussion based mainly on optical data showed that the answer is apparently not (Ozernoy, 1974). The main reason is that the periods observed are much larger than may be expected for a supermassive black hole.

Now basing the data on the radio variability of quasars, I should like to suggest a quite different approach to the problem. The most popular explanation of the radio variability suggested first by Shklovskij and then developed by van der Laan and others is based on the model of an isotropically expanding plasma cloud which contains a 'frozen-in' magnetic field and relativistic electrons radiating by the synchrotron mechanism in the magnetic field which decreases adiabatically during the expansion. This model explains qualitatively rather well some observations, but being confronted with observational data by a quantitative manner encounters a number of difficulties. It turns out that the reason for these difficulties is that one assumes that the thermal plasma is responsible for the dynamics of the expansion. Meanwhile Ozernoy and Ulanovsky (1974) showed that two other kinds of models for variability are possible if the dynamics of the expansion is controlled by the magnetic field or relativistic particles, correspondingly. The best fit of observational data yields the model based on the assumption that the magnetic field controls the dynamics of the expansion. On being applied to some radio variable quasars, the model shows that at radius $R \sim 3 \times 10^{16}$ cm the magnetic field is as large as 10^4–10^6 G, depending on the radial or dipole character of the field.

Such a large magnetic field, in principle, may be obtained from the matter accreting onto a supermassive black hole. However, in the course of an accretion any excess of the magnetic energy density, $H^2/8\pi$, over the matter density energy, ϱc^2, is improbable. Meanwhile for the quasar 3C 279 the observations yield $H^2/8\pi \gg \varrho c^2$. If such an inequality is confirmed by further measurements of the radiovariability as a common case, it should be a new difficulty for the explanation of energy release by means of an accreting black hole.

The value of the magnetic field given above is just of the same order of value as expected from the theory of a magnetoid, i.e. rotating magnetized supermassive body (e.g. Ozernoy and Usov, 1971; 1973a, b). This is a new reason for the preference of a magnetoid as energy source for quasars and galactic nuclei. In principle, the most inner part of such a source may contain a collapsed region of much smaller mass. The above mentioned arguments suggest that the noncollapsed parent body is responsible for the observing manifestations of the quasar and galactic activity.

Note added in proof. Very probable existence of 'proton winds' from the nuclei of Seyfert galaxies offers a new possibility to obtain severe constraints on the collapsed mass as a source of activity (see L. M. Ozernoy, *Astron. Tsirk. USSR*, No. 804, 1 (1973)).

References

Gudzenko, L. I., Ozernoy, L. M., and Chertoprud, V. E.: 1968, *Astron. Zh.* **45**, 492.
Gudzenko, L. I., Ozernoy, L. M., and Chertoprud, V. E.: 1971, *Astron. Zh.* **48**, 472.
Ozernoy, L. M.: 1974, *Proc. First Europ. Astron. Meeting*, Vol. 3, Springer Verlag.
Ozernoy, L. M. and Ulanovskij, L. E.: 1974, *Astron. Zh.* **51**, 8.
Ozernoy, L. M. and Usov, V. V.: 1971, *Astrophys. Space Sci.* **13**, 3.
Ozernoy, L. M. and Usov, V. V.: 1973a, *Astrophys. Letters* **13**, 209.
Ozernoy, L. M. and Usov, V. V.: 1973b, *Astrophys. Space Sci.* **25**, 149.

COMMENT ON ACCRETION AND COMPACT X-RAY SOURCE MODELS

P. BOYNTON, J. DEETER, and D. GEREND
University of Washington, Seattle, U.S.A.

We wish to emphasize further the role of optical observations in studying accretion and X-ray emission processes in the Her X-1 system in particular. Significant visible light variations which are closely correlated with the $\sim 35^d$ X-ray on-off cycle have been observed by us. This variation is largely due to an extremely hot component which is present at maximum light in the $1^d\!.7$ orbital cycle, but only during X-ray inactivity. In much of the remaining part of the 35^d period a secondary minimum is observed. These source properties and possible relevance to a particular accretion model are discussed in a forthcoming paper (Dec. *Astrophys. J.*). Knowledge of this optical 35^d modulation may also enable us to examine the history of this periodicity from 1968 to the present, using the photographic data of Lyutiy *et al.* (1972 preprint); and when applied to current data, can predict the X-ray turn-on time several days prior to that event.

SUMMARY

JOHN ARCHIBALD WHEELER

Joseph Henry Laboratories, Princeton University, U.S.A.

Report what you agree on. Report what you disagree about. Report what should be done to settle the disagreement. These are the famous three instructions of the august patron of several pathbreaking conferences held in Rome. The present symposium is no less historic. However, it is not possible to work out here a similar agreed-upon statement. With 140 of us present, our richly charged symposium is too large for us to undertake that task, and the time available is too short. Also, to our regret, Dennis Sciama could not be here to give us his summary and his answers to the famous three questions. Therefore, in undertaking this review in his place, I can only offer one man's personal perspective.

1. The Symposium in One Sentence

When someone asks us to state the most important result of the symposium in a single sentence, it is difficult to do better than quote the words of our colleague, R. Giacconi, September 7: 'We now have strong evidence in favor of Cyg X-1 being a black hole.'

2. The Search for Black Holes, Neutron Stars, and Gravitational Waves; and Further Predictions about Black Hole Physics

When we are asked for a slightly fuller but still broad brush review, we cannot escape four major points. First, black holes would be beyond our grasp if it were not for the advent of X-ray astronomy. The identification of Cyg X-1 as a black hole is based on the identification of an X-ray and an optical source; on its mass (which in turn depends upon the recent determination of its distance as 2.5 kpc or more); on its compact character; and finally, as especially stressed in the discussions of this morning, on the irregular variations of the Cyg X-1 X-ray intensity on a fractional-second time scale. X-ray astronomy and general relativity have become partners in the investigation of black holes and neutron stars.

Impressive though the evidence is that Cyg X-1 is a black hole, we recall occasions out of the history of physics when what seemed like a discovery turned out to be a mistake. Therefore we are fortunate that there are a few colleagues here and there around the world who look for rival explanations for the observations. No such proposal has been put forward at this meeting; and none of which I have ever heard now remains viable; but not until all such tests have been met and passed will we be able to rank the black hole character of Cyg X-1 as a battle-tested truth.

Second, thanks to X-ray astronomy, neutron stars have shown up for the first time that are married to normal stars in binary systems. As a consequence means are becoming available to determine the mass of an individual neutron star in the married state, as we never could for a neutron star in isolation. Our discussions here make us very mindful of the prospect to obtain in this way masses for so many neutron stars that the critical mass cut-off will stand out, as it did in earlier times for white dwarfs. Then we will have a unique criterion of new precision to distinguish, among compact objects, between neutron stars, below the critical mass, and black holes, above it.

Third, disagreement continues whether any real gravitational waves have yet been detected, but the long term prospects for gravitational-wave astronomy look brighter than ever.

Fourth and last in this broad brush review, theoretical investigations reported here give us more detail than ever on the properties of black holes. We have long known three ways to probe black holes: the pulse of gravitational radiation given out at the time of formation; X-rays given off when matter accretes onto a black hole after its formation; and 'activity', from energy given to its surroundings by a rotating black hole. Little new came out here about the pulse of gravitational radiation at the time of formation. Much was reported that is new and of great interest about accretion. Several most interesting and beautiful aspects of 'activity' were treated. However, we are still some distance from a comprehensive picture of what can happen and what will happen when matter, fields and plasma interact with a 'live' or spinning black hole.

3. The Scope of This Review

Turn now to a more detailed review. Recognize at once that several of the reports presented at this conference were already in themselves excellent reviews. It would be out of place to try to summarize these summaries. Also there were many interesting contributions, so independent and original that they do not fit into any simple pattern. They ranged over topics from the new method of Newman to get solutions of Einstein's equations, and the new solution of Plebanski, both making use of the magic of complex numbers, to the work of Sejnowski on gravitational radiation reaction and absorption, at last making clear the question to which the Bel-Robinson tensor is the answer, and to the report of Grischuk on gravitational radiation of cosmological origin, and to many another fascinating topic, but all together they make a fare too rich to summarize here. If this is agreed, it may be permissible to limit this review to these five topics: sources of gravitational radiation, detectors of gravitational radiation, collapse, X-ray observations, and accretion.

4. Sources of Gravitational Radiation

Sources of gravitational radiation in the early Universe have been imagined, but no such primordial radiation has been conceived so far which is not so attenuated by the expansion of the Universe as to be unobservable.

The thermal gravitational radiation produced in the primordial cosmic fireball should today have the same order of magnitude of energy content as the relict electromagnetic radiation. Moreover, the precise ratio of the two depends in an astrophysically interesting way on the times of decoupling of matter from the two forms of radiation. However, even the most ambitious detector of gravitational radiation ever conceived falls short by many powers of ten of the sensitivity required to determine this ratio.

Ya. B. Zel'dovich has directed our attention afresh to 'fly-by' gravitational radiation, when the two speeding stars are compact, and Nutku and Matzner have an interesting new way to calculate the intensity of this radiation when the velocities are relativistic. Non-relativistic velocities are to be expected in the case of greatest astrophysical interest, the case treated some time ago by Sanders and Spitzer and others, a galactic nucleus that evolves to ever greater compaction. Neutron stars and white dwarfs collide, gas is driven off, and from that gas new stars form still closer to the center. A catastrophe comes closer each day that passes. No one sees any outcome possible but the formation of a giant black hole, a 'nimmer-satt' in the words of Regge. The object grows and grows by accretion of stars. Its mass rises decade by decade, $10\ M_\odot$, $10^2\ M_\odot$, $10^3\ M_\odot$,..., $10^6\ M_\odot$,.... Ultimately it depletes its surroundings. It settles down to a slow terminal stage of growth. Its mass nearly stabilizes at a value that may range anywhere from $10^6\ M_\odot$ to $10^9\ M_\odot$ depending on the mass and compactness of the original galactic nucleus. As Ya. B. Zel'dovich has emphasized, the pulses of gravitational radiation given out in encounters and near encounters between compact star and compact star in the Sanders-Spitzer regime have a frequency spectrum that eludes detection by any of the detectors now in action. Moreover we do not know at what stage the nucleus of our own Galaxy is in the process of building a black hole. Is that process far in the future? Or is it far in the past? Or by some fortunate accident are we in the midst of it today? We know as yet too little on this score to count it as anything but a gamble to build a detector specially designed to respond to these low frequency pulses. When the velocity of fly-by rises from nonrelativistic values to relativistic values we have in the considerations of Matzner and Nutku a new way to estimate the shape and intensity of the spectrum of splash radiation.

Quite another source of low frequency gravitational radiation, waves rather than pulses, was not considered in detail because of shortage of time: the very close binary systems that have received detailed analysis in recent times by Faulkner and Paczynski.

Supernovae, and events like supernovae, have been and remain the potential sources of excitation of a Weber bar of the greatest current interest. Events of this kind, the recent review by Press and Thorne makes clear, should be picked up once a month or so from nearby galaxies with the kind of Weber-bar detector one can hope to see in some years' time. However, it is clear from the discussions in this meeting that we are far from any detailed hydrodynamic analysis of the absolutely central feature of such sources, the time history of the sudden changes in quadrupole moment that take place during this collapse. The greatest variety of scenarios has to be envisaged. The scenario depends upon the mass and angular momentum of the white dwarf core that eventually becomes unstable in the course of its normal astrophysical evolution. Endowed with even a modest initial angular velocity, this core after collapse will be rotating very rapidly. It is natural to think of the collapsed system having under some circumstances the form of a pancake, of neutron matter, which ultimately fragments into two or more neutron stars. One can envisage a so-called 'pursuit and plunge' scenario when initial mass and angular momentum are right. It is natural to believe that there will be the greatest variety of scenarios depending upon the initial mass and angular momentum. Arnett has pointed out to us here some considerations suggesting that the original white dwarf core may never have a mass in excess of 1.4 $M_\odot$. If confirmed, this consideration will be a helpful limitation on the number of cases to be considered. Apart from considerations by Hawking and Gibbons on the minimum number of zeros in the Riemannian curvature of the emitted wave, regarded as a function of time, one knows almost nothing about the details of the shape of the pulse that will be produced in any one of these scenarios. Here is one of the greatest gaps still remaining to be filled if gravitational wave astronomy is to become a useful tool for diagnosing supernova events.

5. Detectors of Gravitational Radiation

Tyson emphasized that detectors of gravitational waves would not exist, or would be rudimentary, if it were not for the pioneering work of Joseph Weber, which he warmly praised. He also pointed out that the detectors attached to a typical Weber bar have improved in sensitivity by two orders of magnitude since Weber's 1970 work; or otherwise stated, the noise level achievable today is two orders of magnitude less than what Weber had to contend with in 1970. He noted that neither Drever nor the Frascati group nor the Munich group nor Braginsky nor the Paris group see anything like what Weber reported in 1969. Either the effect has decreased from then to now by a factor of 100 or more; or Weber's equipment responds to events to which the equipment of others does not respond; or the computer program used to analyze the tapes is at fault; or some other explanation has to be found. The Rochester-Bell Laboratory coincidences with zero time offset between the two tapes show for events of magnitude $\frac{1}{2}kT$ no evidence for anything with a frequency anything like five per day. Garwin's impressive upper limit on the number of gravitational wave pulses, not discussed in detail, increases the evidence reported by these other groups. Fortunately for the future of the subject several of the experimental groups are active in providing other groups with copies of their tapes so that intercomparisons can be made, in the best traditions of experimental physics. One can hope that in a matter of not too many months the numerous experimental groups now at work will be able to give a definitive assessment of 'gravitational radiation' at 1600 cycles.

No one could fail to rejoice at the marvelous new increase in sensitivity of gravitational wave detectors being pioneered by Fairbank at Stanford, Hamilton at Baton Rouge, and the Rome group. When this equipment reaches its ultimate sensitivity, gravitational wave astronomy should be fully established.

Zel'dovich, building on pioneering work of Gerstenstein, Bacalette, de Sabbata, Fortini, Gualdi, Westervelt, Kopvilam and Nagobaroi, described how the gravitational wave travelling in a magnetic field gets converted into an electromagnetic wave, and how a resonant cavity provides such 'conversion' at a great saving on power as compared to free space; went on to point out the impracticability of such a converter under any conditions envisaged to date; but remarked that some new idea may yet be discovered which will make such a converter useful.

Reference was also made in the course of the meeting to the concept of Ruffini, that a charged black hole (if such exists) automatically provides a device to convert incoming gravitational waves to outgoing electromagnetic waves.

6. Collapse

Chandrasekhar presented a systematic review of methods for analyzing the stability of stellar systems, and broadened the methods available for treating the onset of collapse. He noted the special problems which arise in some of these methods in analyzing the stability of the Kerr geometry.

Penrose brought together two ideas that one previously has usually analyzed in disjunction. One is the singularity encountered in models of the Universe itself in the final stages of its gravitational collapse and also at the time of the big bang. The other is the singularity associated with a black hole. He showed how these two apparently different kinds of singularity can be regarded in union. He also pointed out that the concept of 'absolute event horizon' depends on a distinction between world lines which have finite and infinite life. However, no such sharp distinction can be made in a closed Universe of finite volume and finite life. Nevertheless the concept of event horizon under everyday circumstances keeps much of its usefulness. Among topics needing further investigation and elucidation Penrose listed these: (1) Can one assume stationarity (of Newman-Penrose 'constants') for a Kerr black hole? (2) What happens to it when it is subjected to perturbations comparable to the difference $(m-a)$? (3) Can the horizon grow into a singularity?

Zel'dovich pointed out objections to the concept of 'white hole', especially the unrealism of thinking that such an object could remain 'hidden' for a long time and then suddenly erupt. In contrast, even if such objects were allowed, he emphasized, they would have exploded in the very earliest days of the Universe.

These considerations on the inequivalence of past and future in black hole physics remind us of the role of the arrow of time in other parts of physics: in statistical mechanics, in biology, in the force of radiative reaction, in the distinction between retarded and advanced potentials and in cosmology itself. For a closed universe which expands, reaches a maximum content and recontracts, the question has often been raised whether with the 'turning of the tide' the sign of the rate of change of entropy with dynamic time will also change. It is difficult to cite an issue on which there exists today a greater diversity of views. A black hole is an object of very special interest in this connection because in its collapse and in its response to perturbations the one-sidedness in time, the 'friction' that goes with irreversibility plays percentage-wise a larger role than it does in any largescale dynamic process that one can easily name. Surely new insights await us on the connection between the arrow of time and cosmology.

Thanks to Teukolsky we have seen another of the miracles of mathematical physics in the separation of the wave equation for electromagnetic and gravitational waves in the field of force of a Kerr black hole. He, Press, Starobinski and Hartle have told us of the fascinating conclusions they can draw in this way about the interaction of a rotating black hole with its environment. Some of these effects, as for example the ability of a black hole to send out an electromagnetic wave with more energy than the arriving wave (superradiance) are too new for all their astrophysical consequences to have been seen or exploited.

Following Markov's review, 'Global Properties of Black and White Holes', Bardeen summarized the large number of properties of black holes relevant to their observation emphasizing the difficulties of getting black holes with any substantial charge but pointing to the likelihood of black holes with a ratio (angular momentum)/(mass)^2 of the order of 0.95. He pointed out the appearance of a hot blob of radiant matter traveling close to a black hole and shift in spectral lines from it as a tool for studying dynamics in the field of a black hole. He described many of the major features of the accretion disk, a topic of great importance which came up again and again in subsequent sessions of the symposium.

7. Observations

In addition to the evidence on Cyg X-1 presented by Giacconi and Kraft, leading to the conclusion that this object has a mass of 10 solar masses or more and must be a black hole, we heard from Giacconi that there are half a dozen candidates or more for other black holes. Moreover, the chances would appear good that there will be among these objects one or more where the double star system is oriented relative to us in such a way as to give occultation, a piece of good luck not found in Cyg X-1.

Arnett has made us more aware than ever of the hydrodynamic problems of forming a black hole from the precursor star. He points out that the white dwarf core of star models does not exceed approximately 1.4 $M_\odot$. A very much larger mass seems required to produce a black hole of 5 or 10 solar masses. The initial white dwarf core can only hope to reach partway toward this mass requirement if it is endowed with considerable angular momentum, as shown by Ostriker. Therefore a proper hydrodynamic treatment of the collapse of a rotating system would seem necessary if one is to understand how the conditions arise for the creation of a black hole. We have also been reminded in the course of this symposium of other issues of stellar structure and stellar dynamics including not least the fate of a normal star which has ingested a black hole ('cancer of the stomach'). This configuration presents a special case of the general problem of accretion.

8. Accretion

Schwarzmann, speaking for himself and Sunyaev on the search for black holes and the mechanism of luminosity focussed primarily on an isolated black hole in interstellar space, noted that the distinguishing characteristics of the spectrum will be (1) no lines (2) a non-thermal distribution and (3) fluctuations in intensity significant over time intervals as short as 10^{-5} s.

In contrast to the low luminosity of a black hole in interstellar space, a black hole in a double star system will have a much higher luminosity. Schwarzmann suggests a picture for the structure of the accretion disk alternative to that of Bardeen, Pendergast, Rees, Sunyaev, Thorne and others, and reasons that the disk swells, gives way to plasma instability, and breaks into clouds linked by magnetic lines of force. He expects that such a source as Cyg X-1 may be a powerful source of relativistic particles.

9. This Conference and Copernicus

It is difficult to think of any occasion in the history of astrophysics when three stars at once shown more brightly in the sky than our three stars do at this conference today. First is X-ray astronomy. It brings rich information about neutron stars. It begins to speak to us of the first identifiable black hole on the books of science. Second is gravitational-wave astronomy. It has already established upper limits on the flux of gravitational radiation at selected frequencies. At present sensitivity it stands ready to detect the pulses from the next supernova in this galaxy. At the fantastic new levels of sensitivity now being engineered, it promises to pick up signals every few weeks from collapse events in nearby galaxies. Third is black-hole physics. It furnishes the most entrancing applications we have ever seen of Einstein's geometric account of gravitation. It offers for our study, both theoretical and observational, a wealth of fascinating new effects. And it directs our attention more compellingly than ever to the mystery of the final state in gravitational collapse, the greatest challenge in the physics of our generation, and beyond that, to the greatest issue that science has faced in all its history, how the Universe began, and the old question of Leibniz, 'Why is there something rather than nothing?' Surely Copernicus has something to do with those three stars shining in our sky. Surely the spirit of Copernicus will help us with our questions.

INDEX OF AUTHORS

Allen, C. 36
Arnett, W. D. 182
Bardeen, J. M. 132
Bergmann, P. G. 99
Bernat, T. P. 40
Blair, D. G. 40
Bland, R. 37
Bonazzola, S. 39
Börner, G. 186
Boughn, S. P. 40
Boynton, P. 216
Braginsky, V. B. 28, 54
Bregman, J. 181
Butler, D. 181
Carr, B. J. 184
Chandrasekhar, S. 63
Chevreton, M. 39
Deeter, J. 216
Doroshkievich, A. G. 54
Drever, R. W. P. 37
Eardley, D. M. 53
Fairbank, W. M. 40
Faulkner, J. 97
Fortini, P. 59
Fortini-Baroni, L. 59
Gerend, D. 216
Giacconi, R. 147
Gibbons, G. W. 185
Grishchuk, L. P. 54
Gualdi, C. 59
Hamilton, W. O. 40
Hawking, S. W. 184, 185
Herlt, E. 96
Hough, J. 37
John, R. W. 100
Kafka, P. 38
Kemper, E. 181
Koski, A. 181
Kraft, R. P. 181
Lee, D. L. 53
Lee, M. 35
Lessnoff, G. W. 37
Lightman, A. P. 53
MacCallum, M. A. H. 98

Markow, M. A. 106
Matzner, R. A. 16
McAshan, M. S. 40
Meader, D. 52
Melvin, M. A. 101
Misner, C. 3
Newman, E. T. 105
Novikov, I. D. 54
Nutku, Y. 16
Ozernoy, L. M. 214
Paik, H. J. 40
Penrose, R. 82
Persides, S. 95
Plebański, J. 188
Poveda, A. 36
Press, W. H. 93
Pustilnik, L. A. 213
Ramaty, R. 186
Rees, M. J. 194
Rosenblum, A. 102
deSabbata, V. 59
Sato, H. 191
Sazhin, M. V. 54
Sedrakian, D. M. 187
Sejnowski, T. J. 103, 104
Shvartsman, V. F. 183, 213
Starobinsky, A. A. 94
Stephani, H. 96
Stone, R. P. S. 181
Sunyaev, R. A. 193
Taber, R. C. 40
Teukolsky, S. A. 92
Thierry-Mieg, J. 39
Tomimatsu, A. 191
Trautman, A. xi
Tsuruta, S. 186
Tyson, J. A. 17
Wagoner, R. V. 53
Weber, J. 35
Westervelt, P. J. 60
Wheeler, J. A. 217
Will, C. M. 53
Witten, L. 192
Zel'dovich, Ya. B. 54